# A-LEVEL
## AND AS-LEVEL

GUIDES

# GEOGRAPHY

## David Burtenshaw

Longman

## LONGMAN A AND AS - LEVEL REVISE GUIDES

*Series editors*
Geoff Black and Stuart Wall

*Titles available*
Art and Design
Biology
Business Studies
Chemistry
Computer Studies
Economics
English
French
Geography
History
Mathematics
Physics
Sociology

**Longman Group UK Limited,**
*Longman House, Burnt Mill, Harlow,*
*Essex CM20 2JE, England*
*and Associated Companies throughout the world.*

© Longman Group UK Limited 1990

First published 1990
Fourth impression 1993

British Library Cataloguing in Publication Data

Burtenshaw, D. (David), 1939–
    Geography. – (Longman A level revise guides).
    1. Geography
    1. Title
    910

ISBN 0-582-05173-8

Produced by Longman Singapore Publishers Pte Lld
Printed in Singapore

# EDITORS' PREFACE

Longman A Level Revise Guides, written by experienced examiners and teachers, aim to give you the best possible foundation for success in your course. Each book in the series encourages thorough study and a full understanding of the concepts involved, and is designed as a subject companion and study aid to be used throughout the course.

Many candidates at A Level fail to achieve the grades which their ability deserves, owing to such problems as the lack of a structured revision strategy, or unsound examination technique. This series aims to remedy such deficiencies, by encouraging a realistic and disciplined approach in preparing for and taking exams.

The largely self-contained nature of the chapters gives the book a flexibility which you can use to your advantage. After starting with the background to the A, AS Level and Scottish Higher courses and details of the syllabus coverage, you can read all other chapters selectively, in any order appropriate to the stage you have reached in your course.

*Geoff Black and Stuart Wall*

# ACKNOWLEDGEMENTS

I wish to extend my thanks to the following Examination Boards who have granted me permission to use their questions; Associated Examining Board; University of London School Examinations Board; Welsh Joint Education Committee; University of Oxford Examinations Delegacy; Joint Matriculation Board; Oxford and Cambridge Schools Examination Board. In all cases any comments or answers are the sole responsibility of the Author and have not been provided by the Board. The Associated Examining Board deserve my special thanks for making available copies of the teacher guidelines for AS level which have been separately acknowledged in the text.

I wish to acknowledge the permission of the following publishers to print copies of diagrams from their publications; Figs 4.2, 13.1, 13.2, 13.3, 13.6 and 13.7 were taken from *Inequality and Development* by Andrew Reed, and Figures 7.3, 7.4, 7.5 and 7.6 were taken from *Process and Pattern in Physical Geography* by Keith Hilton, all reproduced by kind permission of Unwin Hyman Ltd; Heinemann Educational for Figure 9.2 from G H Dury, *An Introduction to Environmental Systems*; Macmillan Educational for Figure 9.3 from S Nortcliff, *Soils*; J Wiley for Figure 10.4 from S van Valkenburg & C C Held *Europe*; Routledge for Figure 1.1 from D Drakakis Smith *The Third World City*, 1989; *Eurogeo* for Figures 3.4, 3.5 and 3.6 from Eurogeo 2, 1985. Plate 1 was supplied by the author; Plate 2 was printed with the permission of the National Remote Sensing Laboratory and Plates 3 and 4 are used with the permission of Aerofilms. The Ordnance Survey maps are reproduced with the permission of the Controller of Her Majesty's Stationery Office, Crown copyright reserved. Some of the material in Chapter 14 is based on work published in Geography Review 1988 (3) and is reproduced with the permission of Philip Allan Ltd.

My special gratitude is extended to all of those colleagues, friends and family who have assisted me. Murray Thomas my advisor has seen me through many years of examination traumas and I thank him warmly for his encouragement. Thanks also go to my colleagues at Portsmouth; Paul, the two Johns and Janet who have all offered help with their own specialism, Rosemary who drew all of the AEB examination maps and to Christine, Eileen and Michael who prodded me always to devise questions which candidates clearly understood. The preparation of the drafts was a process which owed much to the typing skills of Denise Olway and Sundra Winterbotham, and to the patient editorial expertise of Geoff Black and Stuart Wall.

# CONTENTS

# NAMES AND ADDRESSES OF THE EXAM BOARDS

Associated Examining Board (AEB)
Stag Hill House
Guildford
Surrey GU2 5XJ

University of Cambridge Local Examinations Syndicate (UCLES)
Syndicate Buildings
1 Hills Road
Cambridge CB1 1YB

Joint Matriculation Board (JMB)
Devas St
Manchester M15 6EU

University of London Schools Examination Board (ULSEB)
Stewart House
32 Russell Square
London WC1B 5DN

Northern Ireland Schools Examination Council (NISEC)
Beechill House
42 Beechill Road
Belfast BT8 4RS

Oxford and Cambridge Schools Examination Board (OCSEB)
10 Trumpington Street
Cambridge CB2 1QB

Oxford Delegacy of Local Examinations (ODLE)
Ewert Place
Summertown
Oxford OX2 7BX

Scottish Examination Board (SEB)
Ironmills Road
Dalkeith
Midlothian EH22 1BR

Welsh Joint Education Committee (WJEC)
245 Western Avenue
Cardiff CF5 2YX

# CHAPTER 1

# THE STUDY OF GEOGRAPHY

## GETTING STARTED

You are probably studying at present for one of the eleven A-level courses that are offered in Geography in England, Wales and Northern Ireland, or for the Scottish Higher Certificate, or for the new AS-levels introduced in 1989. The A-level syllabuses have been in existence for many years and every year or two at least one of the major syllabuses, shown in Table 1.1, is the subject of a major radical revision or some minor amendments.

It is probable that, in the light of the GCSE which most of you will have taken recently, further changes will take place. In fact, as this book is written, the Oxford and NISEC syllabuses have just been completely revised for 1990. So do check with your teacher and with your own syllabus in order to be sure that you know exactly what you are to study. You can find the address of your examination board on the opposite page.

At A-level, all Geography syllabuses accord to the **Common Core** which is the Advanced Level equivalent of the National Criteria, which you heard about and read about in your GCSE studies. The view of the core of Geography teaching which has been adopted at A-level, is as follows:

> Geography is a perspective or a dimension of study and a mode of enquiry about aspects of the world rather than a substantive body of factual material. It shifts the emphasis from content to the concepts, skills and techniques which may be considered essential and hence realistically to form the core of the subject. The emphasis is not specifically on detailed content but on a broad framework against which each syllabus is measured. (Common Core Document, 1983)

# ESSENTIAL PRINCIPLES

**CORE GEOGRAPHY**

*" The aims of core geography. "*

The designers of the core for A-level Geography asked 'What ought a student (like you) to have gained as a result of taking an A-level?' As a result, they agreed that seven basic points commanded general agreement among geographers. These were as follows:

- An awareness of certain important ideas in three areas: in physical geography; in human geography; in the interface between physical and human geography.
- An appreciation of the processes of regional differentiation.
- Knowledge derived from a study of a balanced selection of regions and environments, linked with a broad understanding of the complexity and variety of the world in which the student will become a citizen.
- An understanding of the use of a variety of techniques and the ability to apply these appropriately.
- A range of skills and experiences through involvement in a variety of learning activities both within and outside the classroom.
- An awareness of the contribution that geography can make to an understanding of contemporary issues and problems concerning people and the environment.
- A heightened ability to respond to and make judgements about certain aesthetic and moral matters relating to space and place.

| | AEB | Cambridge | London (210) | London 16–19 (219) | Oxford and Cambridge | Oxford | JMB 'B' | JMB 'C' | WJEC | NISEC | Scottish Higher |
|---|---|---|---|---|---|---|---|---|---|---|---|
| **Percentage of A-level marks awarded** | | | | | | | | | | | |
| Essays | 40 | 72 | 55 | – | 50 | 36 | 20 | 40 | 20–40 | | 32.5 |
| Structured | 20 | – | 33 | – | 33 | 36 | | 40 | 20–40 | 35) 35) } 70 | 32.5 |
| Data-response/technique | 20 | 28[a] | 12 | 40 | 17[a] | – | 60 | 20[a] | 30 | | – |
| Decision-making exercise | – | – | – | 15 | – | – | – | – | – | 15 | – |
| Individual field project/ study | 20 | 28[a] | V | 15 | 17[a] | 28[a] | 20[a] | – | 15 | 15 | 25 |
| Coursework | – | – | – | 30 | – | – | 20[a] | 20[a] | | | – |
| Number written examination papers | 3 | 3[a] | 3 | 2 | 6[a] | 2 | 3 | 3 | 3 | 3 | 2 |
| Total examination time (hours) | 7 | 7 | 7.5 | 5 | 9 | 6 | 8.75 | 7.5 | 7.5 | 7.5 | 4.5 |
| **Percentage of** X **AS-level marks awarded** | | | | | | | | | | | |
| Essays | 30 | 25(50)[a] | 40[b] | 64 | *See Cambridge* | 75[c] | – | – | 40[d] | – | – |
| Structured | 40 | – | 40 | – | | – | – | – | – | 70 | – |
| Data response/technique | – | 50 | – | – | | 5 | – | – | 40 | – | – |
| Decision-making exercise | – | – | – | 16 | | – | – | – | – | – | – |
| Field project/study | 30 | 25[a] | 20 | 20 | | 20 | – | – | 20 | 30 | – |
| Number written papers | 1 | 1/2[a] | 2 | 1 | | 2 | – | – | 2 | – | – |
| Total examination time (hours) | 2.5 | – | 3.25 | 3 | | 3.5 | – | – | 3.25 | 3 | – |

| V | voluntary | b | 20% from 'structured report' | d | Physical geography |
|---|---|---|---|---|---|
| a | alternatives exist | c | Human geography | X | COSSEC |

**Table 1.1**

Because the core has been defined in this way it will soon become apparent, if it hasn't already, that your studies might not involve you in learning all that you or your friends consider to be geography. It is impossible to cover *all* the ideas, types of environment and processes that operate upon the world as the home of people. Nevertheless, geographers do feel that it is essential to study human and physical geography and the interface, or the area of overlap, between them. Geography becomes a distinctive subject at A-level by combining the various processes and factors affecting the **natural environment** with those affecting the **human environment**. A physicist may study what happens to air as it is cooled, but only a geographer looks at the way that the results of this process become a hazard. A chemist may know of the complex processes of cation exchange in soils, but it is the geographer who looks at the effects on land use. Sociologists may study deviant behaviour, but the geographer is the one who relates this to localities. Economists examine supply and demand for goods, but the geographer is able to relate supply and demand to the location of the manufacturing process.

## PHYSICAL GEOGRAPHY

Physical geography involves understanding the characteristics of natural environments and the processes involved in the creation, modification and destruction of those environments. Traditionally, physical geography has been subdivided into:

> Aspects of physical geography.

- **Geomorphology** – the study of the form of the land surface;
- **Hydrology** – the study of rivers, lakes and underground water;
- **Oceanography** – the study of seas and oceans (not very common at A-level);
- **Meteorology** – the study of atmospheric processes and phenomena;
- **Climatology** – the study of the long-term state of the weather;
- **Pedology** – the study of soil, its formation and types; and
- **Biogeography** – the study of the distribution of plants and animals.

**The Ecosystem** is sometimes used as a term to describe an alternative approach which looks at the complex interaction of climate, vegetation, soils, animals and people.

## HUMAN GEOGRAPHY

Human geography studies the spatial patterns which arise as a result of the actions of people as decision-makers. Traditionally, human geography has been subdivided into:

> Aspects of human geography.

- **Population geography** – the study of the growth, distribution and movement of population;
- **Settlement geography** – the study of human settlements, their growth functions and patterns;
- **Economic geography** – the study of the distribution of, and factors and processes affecting, economic activities;
- **Historical geography** – the geography of times past and the way the past affects the present. In a way, all Geography is a study of the past – because it has all happened before we study it!
- **Political geography** – the study of political phenomena such as the state and local government and the way that they affect decision-making.

In both physical and human geography it is possible to subdivide these categories, and your syllabuses do just that. On the other hand, the A-level geographer is looking for *links* between the two lists and between parts of each list outlined above. In this way, Geography forms an **interface**.

## REGIONAL GEOGRAPHY

Regional geography describes the way in which many of the relationships have been studied in the past. Most of the generation who are now your chief examiners were served up a diet of *regions* as a form of human geography. The regions were mainly determined by the physical background. Knowledge was encyclopaedic and more appropriate to 'Mastermind' than to modern geography. That form of study is sometimes nicknamed 'capes and bays'. Today, as the third core statement on page 2 notes, the aim is to use regional material to provide examples and illustrations of the processes of human geography. Therefore, in your studies, you may be looking at a variety of **regional scales**, from the small-scale dunes at the mouth of Chichester Harbour, to the medium-scale sphere of influence of a large city, and the global scale of the developing world. You will be studying contrasting regions, such as the Developing and Developed World, capitalist and socialist states, fertile and barren areas, and threatened and stable environments. In your studies you will look at the processes that cause (or interact to cause) these differences, whether they be the result of technological development, population change or government action. The regions which you may study are rigorously laid down in some syllabuses (e.g. JMB 'C' syllabus), while in others there are merely 'suggestions'. However, in all cases there is freedom for your teacher to select the right case study regions for each phenomenon. The region is not the *focus* of your studies, but the relevant *location* for your examples.

> " Consider a variety of scales. "

## THE INTERFACE BETWEEN PHYSICAL AND HUMAN GEOGRAPHY

Human factors influence physical change and physical factors influence human behaviour. For instance, geographers study the ways in which the influence of people on the natural world may have adverse environmental consequences. Sulphur from fossil fuel burning in power stations is the main constituent of the acid rain destroying the forests of Scandinavia. The dredging for offshore sand has taken away the supply of beach sand and accelerated coastal retreat in southern England. A breakwater has starved the beaches of Gold Coast, Queensland, of sand, and so a resort is threatened by a physical phenomenon produced by human action elsewhere. This last example shows that the interface between the human and the physical is two-way. Nevertheless, by understanding the process, action has now been taken to restore the beaches, and people can still surf at Gold Coast. Other interactions are of course more complex, but it is these very interactions which are at the heart of Geography today.

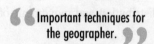

**TECHNIQUES**

To display our understanding of the factors, processes and patterns that we are studying, modern Geography has developed a range of techniques. These techniques help us to understand the problems or issues better because, very often, they give us greater precision. For instance, your A-level syllabus will develop your skills in the use of graphs which were initiated at GCSE. Maps are another essential tool for the analysis and interpretation of data, and all syllabuses place emphasis on a range of maps including, in most syllabuses, the maps of the Ordnance Survey.

> " Important techniques for the geographer. "

Your syllabus will train you in the various techniques concerned with the *collection* of data from the **original (primary) sources** and from the **published (secondary) sources**. The geographer is also trained in the presentation of information, using tables, maps and diagrams. The *analysis* and *interpretation* of information and data at this level will often make use of a number of statistical and graphical techniques. It must be stressed that the geographical techniques are *not* an end in themselves. To draw a map, construct a table or graph, or carry out a statistical procedure, is of little value unless you can *use* that technique to understand a process better, to comprehend a relationship or to appreciate a problem more fully. Therefore most modern Geography syllabuses do expect you to be familiar with a variety of techniques of analysis.

**FIELDWORK**

Geography has always been a field science and all A-levels place emphasis on fieldwork as an essential component of good learning. The 'field' in this sense means the real world and this is the geographers' laboratory; so you should expect to be introduced to aspects of study and enquiry 'in the field'. In this way, classroom teaching and reading are reinforced, and a range of skills and techniques of analysis are taught in a meaningful way. Work may be undertaken in groups or as an individual. It is the latter, the individual geographical study, which now forms a very significant part of the total examination package for most modern syllabuses.

In these various types of fieldwork a range of skills and qualities can be developed so that, besides knowing how to do a task, you develop your own abilities to motivate yourself, to co-operate with others and to work on your own. Geography is developing you as a person and educating you for life, besides training you in certain broad skills of analysis and report writing. Chapter 14 will tell you much more about the various types of project work.

## ISSUES AND PROBLEMS

" Many issues concern the geographer. "

One of the appeals of Geography is that it deals with the real world with all its warts and beauty spots. Besides studying and understanding the processes, you are no doubt aware of many issues and problems. These are often the *result* of the processes, and are of concern to thinking people. The problem of desertification, the issue of the loss of Green Belt land, the loss of industry for one region and the growth of industry in another, are just three examples of the concerns that face us. The list of issues is long, and throughout this book you will find reference to such issues: feeding the world (chapter 4), the exhaustion of fossil fuels (chapter 5), flood hazards (chapter 6), atmospheric pollution (chapter 7), deforestation (chapter 8), dust bowls (chapter 9), deindustrialisation (chapter 10), new towns (chapter 11), and newly industrialising countries (chapter 13). All of these issues have a geographical dimension, but geographers do not claim to have the answers; they merely want to show that Geography is a *dynamic* rather than a *static* subject, looking at physical, social, economic and political change.

## VALUES AND ATTITUDES

In your studies of the processes of change, and the issues resulting from change, you will no doubt discover that different people have different views on the causes of and solutions to, problems such as acid rain or housing shortages. In this way, the study of Geography is a means by which you can develop your *own* values and attitudes and understand why other people hold *their* attitudes and values. By this means, the A-level syllabus is trying to provide you with a deep understanding of the world which you will inhabit once you leave school. It will help you to appreciate the complex nature of the processes of change which will affect you and your friends during your lifetime.

The A-level Geography syllabuses try to fulfil all these aspects of a good geographical education. Much of what has been said here also applies to AS-levels, as you will see below.

## AS-LEVELS

Some of you will be studying for the new Advanced/Supplementary (AS-level) examination, which was first examined in 1989. The AS-level in any subject has the declared objective of 'broadening without diluting academic standards of the curriculum for A-level students'. To achieve this objective the designers of the new syllabuses (who include this author) have had to ensure that the syllabus covers no less than half the content of a conventional A-level. From the statement above we can see that the standards of assessment are to be as exacting as those for A-levels. All that we have said for A-level Geography, in terms of its core and its constituent parts, also holds good for AS-level.

The way that your college or school has approached AS-level will depend very much on the numbers taking the subject, the resources in money, time and staff available, and the syllabus which you study. Most AS-levels have been designed to be taught independently wherever possible, although the boards have all recognised that many centres will probably attempt to combine A- and AS-level teaching. If your AS level is taught separately, then you are getting the best potential deal, but success depends on much more than this. In order to teach the restricted content and yet meet the Common Core requirements, the boards have adopted a variety of strategies. AEB have a set of core topics, which generally relate to the core A-level syllabus, and six optional topics which broaden into distinct areas of study. Both London syllabuses and that of the WJEC are based firmly on parts of the A-level parent courses. In all of these cases the relationship to the parent A-level is clear to see. In the case of Oxford and NISEC, the approach has been to take *one* aspect of Geography (e.g. physical in the case of NISEC) and to relate that part of the discipline to the core. COSSEC has both human and physical courses.

All the AS-level syllabuses have a field study or field project as a major part of the examination. In all cases this work is teacher-marked and then moderated, or checked, by the board. This is an advance on A-levels, where many boards mark the studies and only use your teacher to provide them with supplementary information.

## APPROACHES OF THE SYLLABUS

Each A-level syllabus tries to develop a particular approach to the study of Geography. We can identify the various approaches under a number of headings.

> " The 'three-pronged' study. "

### 1   PHYSICAL–HUMAN DICHOTOMY AND INTEGRATION

Probably the oldest approach that one can detect is what one might term the *three-pronged study* of physical geography, human geography and either regional case studies or techniques of geographical analysis. This approach characterises the London (210), Cambridge and Oxford syllabuses. It usually results in either *separate papers* for physical and human geography (Cambridge/JMB'B') or *separate sections* in the different types of examination paper (London 210). While this approach accepts the traditional dualism (split) of geography, links can still be made between the two halves. As we saw when we looked at the Common Core, there is an *obligation* to study the links between the physical environment and human activity. This integration may come in prescribed area case studies or in techniques questions, or through a form of testing which stresses the interrelationships.

### 2   PEOPLE AND ENVIRONMENT

This approach was once called 'Man and Environment' but this title has been dropped so that all obvious *gender bias* (statements which can and do imply male-dominated processes) is eliminated. The 'People and Environment' theme is, for example, the basis of the AEB syllabus. Even where topics are more obviously 'physical' there is normally a human element to the questions set, so stressing the interrelationships. NISEC has a 'People and Environment' theme to the first half of its syllabus. The London (219), more commonly known as the 16–19 Project, also adopts the same theme through a set of key questions and concepts.

### 3   THE SYSTEMS APPROACH

The Systems Approach is unique to the Oxford and Cambridge Board. A *system* is a set of objects which are organised in such a way that these objects are linked to each other by transfers of material. This approach provides a way of searching for generalisations based on the whole, but through an examination of the parts and how they relate to each other. Therefore the traditional elements of Geography are seen as interacting parts of a total environment. In each environment various processes are dominant, although the outcomes may affect other processes. For example, the hydrological system is one such process, and variations in it can be the result of changes in atmospheric processes, such as higher rainfall totals, and the consequence might be flooding, which has obvious effects on settlement and the functioning of the transport system.

### 4   THE MODERN REGIONAL APPROACH

While the type of regional study which we outlined earlier as 'capes and bays' has now disappeared, that is not to say that there are no regional approaches. For instance, the focus of the JMB'C' syllabus is upon the **Developed World** as exemplified by either the European Community or the USA and Canada, and the **Developing World** as exemplified by Central and South America, Africa or Southern Asia. Within these areas there are a series of **case study regions** within which you are expected to study certain themes. On the whole, the emphasis is strongly slanted towards the human geography of the regions. Therefore the regions become the vehicle for studying aspects such as population geography, agricultural change, industrial location, regional planning and current issues and problems.

In *all* the other syllabuses it is required that you study regions from the local scale to the national and global scales. In this way, you obtain the *real world examples* of concepts, theories, processes and landscapes to illustrate your understanding. Sometimes the focus may be on certain countries to provide a broad range of examples. Brazil and India are favourites in my experience when I have asked questions about development! In the Scottish Higher, the British Isles is the region through which an understanding of the Physical Landscape and Countryside and Urban Settlement is developed. However, major world regions are used as the basis for the study of Land and Water Resources, Aspects of Economic Geography and Aspects of Social Geography.

## 5 OTHER APPROACHES

As Geography develops, so new approaches will emerge. One such development is a theme in the NISEC syllabus: 'Spatial Outcomes Resulting from Culturally Dominated Processes and Systems'. This is the vehicle for organising the study of population, settlement, development and the economic and social organisation of society. It must be stressed that this is a sub-theme rather than the total approach of the syllabus.

A more common approach adopted by many syllabuses is the application of geographical understanding to current issues and problems.

It is widely recognised that the study of geography can help to develop a deeper understanding of many current issues and problems and is therefore valuable for broad educational reasons. This might enable you to gain a firmer grasp of issues involving developing countries, regional planning, use of the Green Belt, the impact of nitrates on water quality, and deindustrialisation, to name but a few. In fact these topics appear in chapters 6, 10, 11 and 12.

**CURRICULUM APPROACHES**

So far, what we have checked on is the approach to the subject based on the underlying theme of the syllabus content. There is another way of looking at what you are studying, and that is the way in which the geography is to be taught. This is called the EBCF, where the letters stand for the Enquiry-Based Curriculum Framework, which is now the approach of the London (219) 16–19 Project.

## THE ENQUIRY-BASED CURRICULUM FRAMEWORK

This approach to Geography is based on a set of key questions and guiding concepts. The 'Route for Enquiry' begins with the *definition* and *description* of a phenomenon, i.e. **what**, and **where**? This leads us to *explain* the phenomenon, i.e. **how**? and to *predict outcomes*, i.e. say **what will happen** if x occurs? Finally, it enables you to **evaluate** and **prescribe**, i.e. say **how should we proceed**? The four 'People and Environment' themes in the London 16–19 Project are structured around this *enquiry-based* framework. They are:

> The 'Route for Enquiry'.

1 The Challenge for Society
2 The Use and Misuse of Natural Resources
3 People–Environment Issues of Global Concern
4 Managing People-Made Environments and Systems

The 'Route for Enquiry' enables you to develop broader skills and techniques that will equip you for life as a citizen and for future study. The skills which you should acquire are:

- communication skills (both spoken and written);
- intellectual skills (thinking, testing, problem-solving);

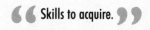

> Skills to acquire.

- practical skills;
- social skills (group work, co-operation and listening to opinions); and
- study skills (organising, using references).

Where the 16–19 Project has led, others have followed, and elements of this enquiry-based approach are now contained in other syllabuses. For example, the advice given for many of the individual project/enquiry papers contains many of the elements of the 'Route for Enquiry'. However, only the London (219) 16–19 Project and NISEC actually go so far as to 'examine' the acquisition of the skills by means of a Decision-Making Exercise (see chapter 2). If one is to crystal-ball gaze, then enquiry-based learning and new assessment styles, such as the Decision-Making Exercise together with individual field project papers, will be of increasing importance. Because of the individual nature of such methods of study, it is inevitable that your teachers will soon be responsible for marking a far higher proportion of your total A-level. They already mark a higher proportion of AS-levels, and so these changes seem to be inevitable.

### AS-level approaches

Everything that we have noted above is applicable to AS-levels. What we have stated about the approaches of individual A-level syllabuses applies in all cases. Inevitably there are exceptions, and these are indicated, as far as possible, in Table 1.1. Obviously any AS-level which is developed in merely human geography or physical geography has to conform to the Common Core, and therefore there will be integrating elements to bridge the dichotomy.

## CONCLUSION

I hope that you have been able to read this chapter quickly because you already know about *your* syllabus and its approach to Geography as a discipline. Without reproducing each syllabus in full, it is impossible to tell you more in this brief introduction. If you want to know more, then you should obtain a copy of your syllabus for the year in which you sit the examination (for January examinations that is the preceding year). Get this from your teacher or from the Examination Boards whose addresses appear on page vii. If you are studying alone, then it is advisable to consult the board of your choice early in your studies; they are there to help you.

With clear understanding of the needs of your syllabus, it will be much easier for you to structure your studies and your revision so that you obtain the best possible result from your efforts.

# CHAPTER 2

# STUDYING, REVISING AND THE EXAMINATIONS

## STUDY AIDS CHECKLIST

## STUDYING GEOGRAPHY – COURSEWORK

## REVISING

## ESSAY PAPERS

## SHORT STRUCTURED QUESTIONS

## DATA RESPONSE/ TECHNIQUES

## GETTING STARTED

You are aware, no doubt, how much there is to study at A- and AS-level. Geography has always had a reputation for being a very full subject to study in the time available. You cannot therefore rely on a last-minute rush of study. In fact, all A-levels depend on a gradual acquisition of knowledge, understanding and skills, and therefore you should try to adopt a steady pattern of study even if your course is assessed only in the last term of your studies.

Some rules for studying during the course:

- Always keep your notes and handouts ordered and up to date.
- Do all required work on time – there is a reason for the timetable of work.
- Approach your teacher if you do not understand bits of the course – always be ready to ask questions.
- Make every effort to attend the fieldwork elements because they are the laboratory work for the geographer.
- Always read around a topic, using the books which you have been given, and the journals noted in the Study Aids Checklist below.
- Keep press cuttings with the relevant section of your notes.
- Keep especially clear notes and handouts which explain statistical techniques – have a note telling you where there is a textbook working of a technique.
- Do enter into the classroom discussions – these improve your communication skills and enable you to see whether you understand the topics.
- Do some Geography work outside of the classroom every week.

# ESSENTIAL PRINCIPLES

As you study, you should make sure that you build up the following aids to learning and understanding.

❝ Ask yourself these questions. ❞

- **Have you got your notes from lessons, handouts and any extra notes organised in the order of the syllabus?**

- **Have you got an example of each of the concepts, theories, models and processes in your notes?**

- **Do your notes tell you where to find out more information in books, journals and newspapers?**

- **Have you got easy access to a copy of a textbook which covers the basic elements of your syllabus?**   There are textbooks which cover the whole of most syllabuses, for example, B. Knapp's *Systematic Geography* (Unwin Hyman, 1986), but maybe you have a class text which covers half or some of the subject matter, for example, M. Bradford and A. Kent, *Human Geography: Theories and Their Application* (Oxford, 1977), or A. Goudie *The Nature of the Environment* (Blackwell, 1984). If the school or college cannot provide you with a textbook, perhaps you should buy one. You can always sell it to someone in the next year when you pass!

- **Are you aware of the books which the examination board suggest for your study?**   Most boards issue book lists, but do remember that they are only suggestions to help you and your teacher with your studies. The lists normally include books for the teacher or lecturer which those of you aspiring to the Special paper, or to special university entrance examinations, should read. Sometimes the lists will contain shorter, easily readable texts on individual sections of the syllabus, for example, D. Burtenshaw, *Cities and Towns* (Unwin Hyman, 1986) or A. Reed, *Inequality and Development* (Unwin Hyman, 1987), M. D. Newson, *Hydrology: Measurement and Application* (Macmillan, 1979) and A. Kirby and D. Lambert *Land Use and Development* (Longman, 1985). Be aware of these when doing essays as a part of your studies. The London (219) 16–19 Project has spawned its own series of short study unit texts: for example, *Coastal Management: a case study of Barton-on-Sea* (Longman, 1984). Obviously, these are a *must* for the 16–19 syllabus but they may be used by teachers and students of other syllabuses.

- **Where can I find the books which I need?**   Are they held for allocation in the Geography Room or in the School Library? If the school cannot afford to purchase them or if you are a private student, why not arrange with your fellow students to borrow them from your local public library? If the public library does not have a book, then you may ask for it to be obtained on *inter-library loan* from another library. You could all consult it before it is returned.

❝ Don't forget journals. ❞

- **Are you aware of the journals which contain relevant material to help you with your studies?**   Journals can provide more up-to-date examples and, especially, ideas for potential individual projects. *Geography Review* is a journal which your school/college can obtain at a special discount rate. It is written for *you* by people who know what help *you* need. The six editions a year even contain worked answers just like the later chapters in this book. *The Geographical Magazine*, the monthly magazine of the Royal Geographical Society, also contains articles on a wide range of topics relevant to your studies. This can also be bought at a special educational discount. *Geography*, the journal of the Geographical Association, the teachers' association, might also be of help on occasions. *Geofile* is a further useful source of data.

❝ Newspaper cuttings are an important resource. ❞

- **Have you kept a file of relevant newspaper cuttings?**   The quality press, that is, *The Times*, the *Guardian*, the *Independent*, the *Financial Times*, *The Scotsman* and the *Daily Telegraph* all contain articles of relevance. Some of the longer articles even appear as the data or stimulus for questions, and decision-making exercises invariably involve the use of newspaper articles. Local press reports can provide ideas for individual projects, as well as excellent examples of processes. For instance, the flash-flood which hit Rottingdean on 8 October 1987 produced several weeks of commentary and photographs. The *Financial Times* has frequent supplements on all kinds of topics, regions and countries. Other papers listed above do also produce special supplements.

■ **Do you possess other study aids?** Geographers need coloured pencils for maps (much better than the ugly felt-tip pen), rulers, etc. for diagrams, graph paper and a calculator for the essential data processing.

### For the one-year student

Both A-level and AS-level are supposed to be studied over a period of two years. However, some of you may be approaching a one-year course, or even a short course, prior to repeat examinations in the winter (Cambridge, and London (210) both have a winter examination, in November for the first, and January for the other). The one-year student has to be more organised and dedicated. The checklist needs to be observed from the outset and you will have to decide how to limit your studying so as to develop the depth of understanding required in the time available. Doing everything superficially is not good advice; you need to focus on enough syllabus content to ensure that you can attempt *all* the examination. Some syllabuses, for example London 16–19 and WJEC, make such an exercise difficult by design. As a one-year student, you will have to be more aware of the coursework and/or individual project submission dates – normally 1 April or 1 May for most project papers. If you are studying on your own, contact the board's Geography Subject Officer and obtain the relevant details. (The addresses are on page vii.)

## STUDYING GEOGRAPHY – COURSEWORK

Coursework is not a synonym for homework. At A- and AS-levels in the 1990s a part of your grade will be based on work completed during the two-year course. A major part of this move to coursework is the *individual project*, which we will cover in chapter 14. In addition more syllabuses are expecting to assess your everyday work. This assessment might lie in the board's hands but, increasingly, coursework is seen as being more appropriate to teacher assessment with board moderation (i.e. checking).

In the London (219) 16–19 Project, there is a greater emphasis on teacher assessment than in any other syllabus. Your school or college is responsible for 30 per cent of the total marks which are gained from the Option and Special Option Modules. Teacher assessment of your work is also a possibility in both the JMB syllabuses, although it is an alternative to other papers. WJEC is a third board to introduce teacher assessment of your work.

At AS-level, the field project/essay is marked by your teacher in every case.

## REVISING

If you have developed the right study skills with the help of your teacher, revision becomes less daunting. Revision is easy if you have been organised throughout your studies. Revision should be a gradual build-up to the papers and it should be balanced. Geography is one of your subjects – do not neglect the others or vice versa. Similarly, make sure that your revision covers enough of the syllabus areas and/or themes to give you the best chance with unseen examinations.

### REVISION GUIDELINES

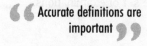
Accurate definitions are important

1  Make sure that you know the meaning of terms, concepts, theories. Here are some basic terms; I have chosen one for each of the subsequent chapters and the definitions of each of these may be found there:

■ Crude death rate (ch. 3)    ■ Illuviation (ch. 9)
■ Stock resource (ch. 4)    ■ Extensive agriculture (ch. 10)
■ Primary energy (ch. 5)    ■ Counter-urbanisation (ch. 11)
■ Clints and grykes (ch. 6)    ■ Dependency (ch. 12)
■ Dry adiabatic lapse rate (ch. 7)    ■ Urban Development Corporation (ch. 13)
■ Heterotroph (ch. 8)    ■ Cycle of deprivation (ch. 14)

Recently, various dictionaries of geography have appeared and these do provide you with clear, understandable definitions:

J. Small and M. E. Witherick, *A Modern Dictionary of Geography*, (Edward Arnold, 1989). Clear and easy to follow – authors are chief examiners and write for your level.
B. Goodall, *The Penguin Dictionary of Physical Geography*, (Penguin, 1984).
J. Whittow, *The Penguin Dictionary of Human Geography*, (Penguin, 1984). More complex and more complete – of value for university students.
R. J. Johnston, *The Dictionary of Human Geography*, 2nd edn, (Blackwell, 1985).
A. Goudie, *An Encyclopoedic Dictionary of Physical Geography*, (Blackwell, 1985). A more advanced dictionary for university use, but valuable for the S-level candidate.

2　If there are diagrams or models which can be used, do you know them and can you draw them? You should have a memory bank of relevant diagrams, such as Smith's 1966 space-cost curve (Fig. 10.11) used in industrial location to illustrate spatial margins, least-cost location and optimal location (see chapter 10).

3　You must be prepared to comprehend the relevant statistical tests and illustrate what they mean in different contexts. For most boards, there is no need to learn formulae or to complete calculations (JMB and London do require working for some questions). Many boards, such as Oxford, issue formulae guidelines and you should be aware of these.

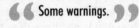
**Practise writing answer plans in 3 to 5 minutes.**

4　Make sure that you are able to work within the prescribed length of time. If you have an essay paper of four questions in three hours, then you should have attempted to write essays within 40–45 minutes. Similar practice attempts can be made for questions from other past papers, during the last two months before the examination.

5　Try to ensure that you have looked at Ordnance Survey maps, and photographs, during the revision. (The Oxford Delegacy does not require Ordnance Survey mapwork.) Their interpretation requires particular skills which are very geographical and need practice. On the whole, you cannot learn mapwork and, therefore, you need practice at looking rather than learning for these skills.

## REVISION WARNINGS

**Some warnings.**

- DO NOT ignore too many parts of your studies in order to cut down on revision. In these days of integrated questions, you might be forced to use your recollection of a section you ignored in order to answer part of a question on a topic you regard as one of your strengths. Techniques questions might require you to attempt parts which you have 'ignored'.

- DO NOT pay too much attention to essays or other questions done for homework. These are normally past questions and a verbatim repeat of a question, even if the theme recurs, is an extremely rare event. New questions will have a different slant and an answer to the previous question on the same topic will generally score poorly.

- DO NOT FORGET TO re-read any project work and your field-work notes because this work can provide you with excellent examples.

- DO NOT leave revision for the second paper until the first is over. This is particularly dangerous on London (210) and AEB where the question format varies although it is the whole subject matter that is being tested. However, you can normally discount the possibility of an identical topic occurring twice in one set of examinations.

# THE EXAMINATION PAPERS

Because Geography is attempting to test your acquisition of differing skills, it is understandable that the papers and the questions are very varied. Some papers may consist of only one type of question; other papers may have mixed questions.

## ESSAY PAPERS

The traditional examination at this level has been the written essay paper and all boards make use of this as a part of their assessment. Of course, clear structure is important in any essay answer, perhaps with separate paragraphs for each main point. Remember to have an introduction, where you explain to the examiner how you are going to approach the problem. Remember also to have a conclusion, where you review your findings in the light of the question actually set.

AEB has one essay paper (40 per cent) with four answers required from sixteen questions, one from each of the four sections of the syllabus. London (210) has two essay papers which comprise 55 per cent of the total mark for the examination. The papers require that you answer four essays from eighteen and two from ten. Essays in the Cambridge paper can gain 72 per cent of the total marks; each paper requires a response to three from at least eleven questions. Oxford and Cambridge have three essay papers which expect two from four, two from twelve and two from ten to be answered. JMB 'C' relies less on an essay format, with only 40 per cent of the marks from four essay parts of two papers. The Scottish Higher requires essays on both of its papers.

Here are some *essay titles* which should give you an idea of the types of essay that are set. Beneath each one I have tried to suggest what the essay type is and what it is testing.

### Question 1

For any one country or region, explain why the relative balance between population and resources is not an optimum one.

(AEB, June 1987: *25 marks*)

> Be aware of the different types of essay.

This is the *single-topic* question. It wishes you to show that you have some idea of the concepts of *optimum population* and *population and resource balance*. You are to demonstrate your understanding of the concepts by *explaining* why. You also have to apply your theoretical knowledge and understanding to a case study region or country; obviously, if there is no supporting case study evidence, then the mark will be restricted.

### Question 2

a) What do you understand by the term 'growth pole'?                                    (*5 marks*)

b) What circumstances would persuade a government to set about creating
   a 'growth pole'?                                                                       (*10 marks*)

c) How might a 'growth pole' be created?                                                 (*10 marks*)
                                                                                         (London)

In this case, the essay has been partly *structured* and the balance of the answer is implied by the balance of marks. The first part is testing your knowledge of a particular concept, and the second part is testing your understanding of the concept and its application. In the final part, you are given more freedom to write about how a 'growth pole' might be created, although some reference to an actual case would help considerably.

### Question 3

Assess the relative importance of climate and parent material as factors affecting the character and properties of the soil.

(AEB: *25 marks*)

This is an *evaluative* essay question; another format begins 'To what extent . . .?' and another 'How far do you consider . . .?'. In this case, you have to examine both climate and parent material in relation to the character of the soil as well as its properties. In order to assess, you have to ensure that your conclusion does actually weigh up the merits of each factor.

The other two formats also require some attempt to evaluate. For instance, if you are asked 'To what extent does transport control the structure of a city?', you would be expected to deal with factors other than transport as well as with transport itself. In your conclusion you could state whether your analysis leads you to believe that transport or some other factor(s) controls the structure of a city.

### Question 4

Prepare an outline paper for an incoming Minister of Transport on the problem of providing public transport in rural areas and suggest possible solutions to this problem.

(AEB, June 1988: *25 marks*)

This is a *role-playing* question which asks you to demonstrate the relevance of your geographical knowledge. It requires you to write an essay which demonstrates an understanding of the problems of rural areas and then to use your own attitudes and values, with the possible help of real cases, to suggest solutions. Although the question does not say 'with examples', it is in your obvious interest to provide them. Examiners would not expect a Civil Service style of writing; they just want to see a well-constructed case. The structure of the essay is in the question – (1) problems and (2) solutions – but the balance is not clear, so assume that it is relatively even. The examiners would expect more knowledge of problems than solutions.

The key requests in essay questions:

| | |
|---|---|
| *Describe* | do just that – do not explain. |
| *Briefly describe* | be concise and straightforward – do not explain. |
| *Outline* | do just that – be concise and straightforward. |
| *Illustrate* | describe with examples – usually linked to explanation. |
| *Explain (or Account for)* | make sure that you give reasons or causes. Description will not suffice. |
| *Justify* | make sure that you give your reasons and even your opinions. |
| *How do?* | another form of explanation. |
| *To what extent?* | evaluative – there are several possible explanations – give these and try to say which you tend to favour. |
| *Assess* | evaluative rules apply again here. |
| *How far?* | evaluative rules apply – expects agreement or alternative explanation. |
| *Compare* | find the common elements – some contrast might be implicit. |
| *Contrast* | what are the differences? – often linked with compare, so try to keep the sections separate. |
| *Discuss* | the academics' favourite! In other words, outline and explain a range of factors. Sometimes the question will imply evaluation as well. |
| *Classify* | make sure that there is a classification in your answer. |
| *Consider the view that* | evaluative – possibly asking you to hold a particular attitude. |
| *With reference to specific examples* | make sure that the examples are just that and not vague 'as in London' or 'as on the Dorset coast'. |
| *Why is it?* | another question expecting logical explanation. |

> **Know what each term is asking you to do.**

## HOW GEOGRAPHY ESSAYS ARE MARKED

As more syllabuses specify the assessment objectives in terms of knowledge, understanding, skills and values, so the examiner has to take account of their relative weight in marking your essays. The marks awarded for each essay or part of an essay are indicated on the paper. If an essay is being marked out of 25, then it probably has a banded mark scheme where the examiner is looking for certain standards to be attained in order to place the essay in that band. An example of a banded marks scheme follows. Here, there are descriptions of the types of attainment expected of every set of marks. Sometimes the essay will not fall into any one category and so the examiner will exercise professional judgement and, no doubt, explain the decision at the end of the answer. In addition, some guidance on the types of points which could be developed is given, and finally some rules for those who do not follow the question are laid down.

### Question 5

For any **one** country or region, explain why the relative balance between population and resources is rarely an optimum one. *(25 marks)*

### Criteria bands

21–25   An excellent command of the information which supports a good case based on the selected area. Do not expect there to be perfection in terms of data but there will be figures to suggest that the relationship is not optimal. There is balance between the two components of population and resources and the essay will include an historical element. Not much that an 18-year-old could add.

16–20   Still a good coverage but the essay is probably rather unbalanced between the two components. Supporting evidence more patchy and occasionally inaccurate. Historical element less obvious and more by accident.

> These bands will give you an idea of what you have to do to score well.

10–15   The average answer where the ideas are known but expressed in a fashion that suggests learning by rote or a random recall of relevant points. Historical element thin or accidental. More unbalanced between two components. Examples are correct but generalised and slightly exaggerated. Vague knowledge of optimum population. Superficial use of country or region. Must say 'why' the balance is rarely optimum.

5–9   Aware of the question's intentions but unable to provide a convincing argument. Probably all about one or other component rather than both. Never states what optimum population is even by implication in the text. Disorganised ramble through relevant material but with little support from area. A good non-place case.

0–4   Does not state area and provides no support to the case which is theoretical. A few ideas in a final rush. Just about one component without an appreciation of the other – resources probably. No country or region.

### Factual information

Relationship is rarely optimum because:

1   The pattern of resources changes over time;
2   Resources vary in output over time: for example, agricultural, mineral discoveries;
3   Not all resources are material – some are financial and enable imports to be met;
4   Optimum is a theoretical position;
5   Population structure varies through time and the optimum rarely achieved;
6   National statistics conceal regional variations within a country.

### Marking rules

If no country or region cited, then 9 is the maximum. If continental (e.g. Africa), then maximum is 9. If country is only cited in a frequent 'as-in-Nepal' style, then 14 is the maximum.

Finally, it is worth repeating that examining is becoming increasingly sophisticated. The Grid (Table 2.1 pp. 15–16) is an attempt based on some syllabus proposals at A- and AS-levels to show how the various marks fit broad descriptions of your essays, no matter what the essay title may be. The mark bands give an indication of what the examiner is looking for under some of the headings for each answer. Read across the top row if you wish to achieve the best marks for each heading. However, if you are scoring less well, then look at your essay in relation to the *row above* your current mark; this will give an indication of what you have to do to score better and, hopefully, to improve your final grade.

| Marks out of 20 | With regard to knowledge and understanding | With regard to logic and organisation | With regard to support, examples | With regard to technique | With regard to instructions | Other points |
|---|---|---|---|---|---|---|
| 17–20 | Thorough understanding of concepts, theories, principles. Uses terms, jargon correctly. Accurate information. | Very logical. Well organised. Able to analyse, permutate and re-combine. Creative argument. A very good essay. | Examples fully developed. Maps and diagrams used fully. Accurate locations, regions. | Appreciates techniques and uses them. | Homes in on demands. Evaluates well, describes fully, explains completely. | Perfect for 18-year-old in time. We could not do better. |
| 13–16 | Relevant, accurate, good knowledge. Concepts, theories and principles understood but some missed at lower end. | Logic a little slipshod. Occasional last-minute thought. Comparison fades into parts. Theme not appreciated in total. | Examples relevant – may dominate knowledge of theories. Can represent ideas. | Good techniques. Diagrams used. Knows technique but lacks appreciation. | Evaluates by implication. Explanation is reasonably full. Description complete. Balanced discussion. | A competent performance. |

| Marks out of 20 | With regard to knowledge and understanding | With regard to logic and organisation | With regard to support, examples | With regard to technique | With regard to instructions | Other points |
|---|---|---|---|---|---|---|
| 8–12 | Basic ideas and concepts. Some facts, etc. misunderstood. Normally basic textbook material – regurgitates notes or a previous question. Can be short on content. | Partial grasp of argument. Logic tends to become whatever thought of next. Poor expression. Shotgun technique of undeveloped points. | Examples increasingly sketchy, 'as in . . .'. Some support not really ideal. | Diagrams adequate. Techniques known but not really understood. | Lacks evaluation. Straightforward description. Explanation incomplete. Discussion unbalanced. | Represents the average candidate. Goes for breadth rather than depth. |
| 4–7 | Shows a little knowledge based on study. Simplistic grasp of concepts and ideas. Omission and mis- understandings obvious. Irrelevance increases. Sensationalism and bigotry. | Lacks ability to organise material or to develop ideas. Drifts from question. | Possibly no examples or very intermittent examples. Support may be wrong. | Poor comprehension of diagrams. Techniques not understood and sporadic. | Low level description. Erroneous explanation. Merely description masquerading for explanation. | An incomplete answer; note form; last part to a last question. |
| 1–3 | Poor understanding of simple concepts and ideas. Half-truths and irrelevance. | Little idea of how to make a case. Misses point. Random assemblage. | Probably no support or misplaced. Sensational/ bigoted. | Diagrams wrong, incomplete or rare. | Little attempt to follow instructions. | Could be a last gasp single paragraph from an otherwise competent paper. |

Table 2.1 A guide to marking essays (the bands will vary if the marks total is different, e.g. 25)

## SHORT STRUCTURED QUESTIONS

These are the standard method of testing your understanding of the basic models, theories and concepts. In some cases, (AEB, Oxford and Cambridge, and London (210)), they are placed in a separate paper. Likewise, structured questions appear in two JMB 'C' papers and constitute 40 per cent of the final mark. Scottish Higher also uses short questions.

The question below is a typical structured question. You might attempt to answer it if you wish to test your geomorphological or, more precisely, hydrological knowledge.

### Question 6

Study the example of a storm hydrograph illustrated below.

> Structured questions are very popular.

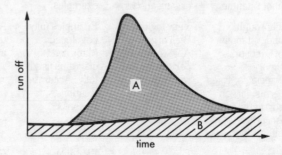

Fig. 2.1

a) Name the flow components labelled A and B.

A _____

B _____ (3 marks)

b) Explain why the flow labelled B has a more even regime than the flow labelled A.

(6 marks)

c) In what ways and for what reasons would extensive urbanisation affect the storm hydrograph?

(*6 marks*)

(AEB, sample AS, 1988)

This type of question usually leaves gaps for your answer. How is this question testing you?

Part a) relates back to the diagram which you have been asked to use to prompt your reactions. It wants you to show knowledge of the components of run-off. One-word answers, using terms, will suffice here.

Part b) is testing your understanding or comprehension of the graph. It wants you to give the reasons in as short a way as you possibly can.

Part c) develops the theme of the hydrograph to include the 'People and Environment' theme by focusing on the role of urbanisation. It asks for knowledge in terms of the ways, and understanding in terms of the reasons.

This question was designed to be answered in 15 minutes. You can see that you do need to have the facts at your fingertips so that you can demonstrate your understanding within as short a time as possible.

In this type of paper you should assume that you are supposed to write within any lines which might be provided on the page.

### Rules for answering structured questions

- The marks for each part are a guide to the time. Part (a) here should take little time, i.e. up to 3 minutes, (b) 6 minutes, and (c) 6 minutes. This would vary according to the examination being taken.

- Do not be put off if you cannot do one of the parts, especially the first. Sometimes, merely reading the other parts jogs your memory.

**Some useful rules.**

- Always attempt the correct number of questions. *Do not gamble with too few.* If you answer too many, the examiner will mark all and then cancel the lowest mark. This strategy of doing more *does not* normally gain good marks because you have possibly spent too little time on each answer.

- Pay careful attention to the question. If they want *two* reasons, then make sure that there are *two*. Also, make sure that your two responses are not variants of the same point.

- If examples are asked for, then give them. If it says *specific*, then make sure that the example is that, for example a specific regional example of a cycle of resource development is *not* coal in the UK but coal in South Wales or even coal in the Rhondda.

- If you are asked to draw a diagram, model or soil profile, make sure that it is labelled, with a clear key. For graphs, make sure you have labelled your axes.

- If you have to cross out and answer elsewhere in the answer book, or on a separate sheet, then tell the examiner where to find the answer, for example, 'answer continues on page x' or 'please see extra sheet'. If extra sheets are used, fill one before starting another. However, most examiners do find extra sheets a nuisance and there is a small chance that they might get lost. Justifiable extra sheets, etc., will always be marked but the best rule is *not to use* them.

In the following chapters you will find an example of a structured question for each area of study. In some cases, these will have candidate answers and in others you could test yourself.

### DATA RESPONSE/ TECHNIQUES

The third type of assessment takes the form of presenting you with previously unseen geographical data and using this to test your ability to understand techniques and to interpret information. The information may come in a variety of forms which will be illustrated throughout this book, though the list below contains the most common forms. The data may be presented singly or in combination.

### Types of data

- Tables – with associated statistical texts, for example correlation.
- Graphs and thematic maps.

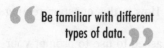

- a) Photographs – of landforms and townscapes.
- b) Aerial photographs – both vertical and oblique.
- c) Satellite photographs – mainly used for meteorology to date, but be aware of satellite images.
- d) Sketches of landforms.
■ Ordnance Survey maps – 1 : 50,000 and 1 : 25,000 from the various series.
■ Land use maps – the 1960s survey is now dated, so use is limited.
■ Soil maps and geology maps – might be used.
■ Stimulus-response – for example the storm hydrograph material in the previous section is of this type.
■ Statistics – these are normally given with the result of the test so that you can interpret the data. Most boards do not expect you to learn formulae or to make calculations.
■ Written text can be a form of data which can be interpreted.

The questions which are set usually contain a combination of materials on a theme, so that you can exhibit your skills of analysis and interpretation alongside the knowledge you have gained from your studies. The question below contains two-line graphs, a bar graph, and two tables. (Table 2.2 is really brief data to help in the answer.)

| | GNP PER CAPITA | | POPULATION | | URBANISATION | | |
|---|---|---|---|---|---|---|---|
| | $ (1981) | % annual growth (1960–80) | millions (1981) | projected % annual growth (1980–2000) | % population urban (1981) | % in cities > 0.5 m | % in largest city (1980) |
| Mexico | 2 250 | 3.8 | 71.2 | 2.6 | 67 | 44 | 32 |
| Average of 19 industrial countries | 11 120 | 3.4 | — | 0.7 | 78 | 55 | 18 |
| Mexico City – Population: 1950 3.2 million 1980 16.0 million 2000 31.0 million (projected) | | | | | | | |

**Table 2.2**

a) **Flows of Capital 1979–80**

  i) Into Mexico:
    $664.9 million
  ii) Out of Mexico:
    Interest     $401 million
    Profits      $284 million
    Licence fees $298 million
    Other      $39.6 million

b) **Major funders of external debt**

  USA    City Bank
          Bank of America
          Chase Manhattan Bank
          Chemical Bank
          Morgan Guaranty Trust
          Manufacturing Hanover
  Others  Gulf International Bank
          Union of Franco-Arab Banks

c) **The National Debt**

In 1983 petroleum exports from Mexico were totally pledged to the USA in return for a loan of $1 000 million.
The total 1983 debt interest and repayments in 1983 were $14 000 million.

**Table 2.3**

## Question 7

Figs 2.2, 2.3 and 2.4 show respectively (i) foreign investments in Mexico between 1940 and 1978 by economic sector, (ii) the country of origin of foreign investments and (iii) the changing pattern of foreign manufacturing investment in 1950, 1960 and 1978.

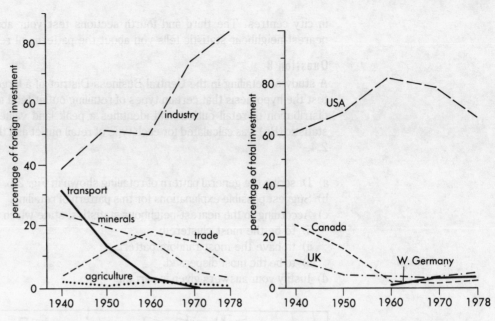

Fig. 2.2  Foreign investments in Mexico, 1940–78

Fig. 2.3  Country of origin of foreign investment, 1940–78

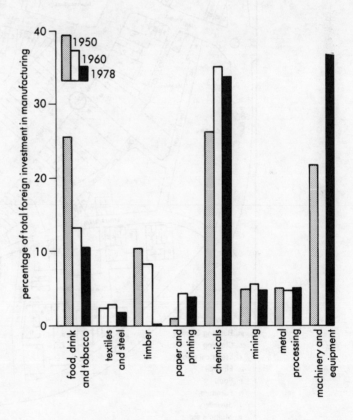

Fig. 2.4 Changing pattern of
foreign investment, 1950–78

a) Describe the changing pattern of foreign investment in Mexico over the period
   1950–78.                                                                      (*10 marks*)

b) Table 2.2 shows some of the economic, demographic and urban growth characteristics
   of Mexico. Table 2.3 shows other aspects of Mexico's trade and economy.

   Using this tabulated data and Figs 2.2, 2.3 and 2.4, identify and examine those
   characteristics which suggest that Mexico is a developing country.            (*15 marks*)

                                                                        (AEB, June 1987)

The first part is testing your ability to *use* the data to aid your understanding. Having done
that, the question proceeds to see whether the information can be combined with your
previously learned material on the nature of a developing country's economy.

   The next question provides a map (Fig. 2.5) and the results of a set of nearest-neighbour
statistics (Table 2.4) for your use. In the first part, it is asking for the simple skill of being
able to read a map. You are then expected to relate this to your understanding of retailing

in city centres. The third and fourth sections test your ability to understand what the nearest-neighbour statistic tells you about the patterns of retailing.

### Question 8

A study of retailing in the Central Business District of a large UK city was carried out to test the hypothesis that certain types of retailing outlet tend to cluster. Fig. 2.5 shows the distribution of retail outlets and identifies a peak land value point. A nearest-neighbour statistic (Rn) was calculated for each type of retail outlet and the results are shown in Table 2.4.

a) Describe the general pattern of retailing shown in Fig. 2.5.       (*8 marks*)

b) Suggest possible explanations for this pattern of retailing.       (*10 marks*)

c) According to the nearest-neighbour statistics, state which type of retail outlet appears:
   i) to be the most clustered;
   ii) to have the most random pattern;
   iii) to be the most dispersed.       (*3 marks*)

d) Justify your answers given in part c).       (*4 marks*)

(AEB, November 1987)

B Bank
BS Building Society
C Clothing
E Electrical goods
Fin Finance
F Food
Fu Furniture
J Jewellery
M Multiple store
P/R Pub./Restaurant
S Shoes
O Other retail outlets

Pedestrian

(X) Peak land-value point

100 m

**Fig. 2.5**

**Table 2.4** Nearest neighbour statistics for selected types of retailing

| Type of retailing | Number of outlets | Rn statistic |
|---|---|---|
| Clothing | 45 | 0.55 |
| Shoes | 10 | 0.51 |
| Furniture | 9 | 0.8 |
| Food | 9 | 1.8 |
| Electrical goods | 11 | 0.56 |
| Jewellery | 7 | 0.61 |
| Multiple stores | 8 | 0.47 |
| Banks | 5 | 0.8 |
| Finance | 8 | 0.2 |
| Building societies | 10 | 0.26 |

Photographs are often used to see if you can interpret what you see, and this is nearest to being a test of your ability to use your eyes and understand landscapes. Photographs may be in colour or black and white. They are increasingly in colour because most modern photographs are in colour and our world is a world of colour. In an exam question you can expect a panoramic photograph such as the one reproduced in black and white in chapter 6 (Fig. 6.20) which expects you to identify landforms and explain the processes operating on those landforms. Alternatively, you may be given an oblique aerial photograph, in this case of two different types of urban scenery (chapter 11, Figs 11.25 and 11.26) which were used in conjunction with an Ordnance Survey map. In an examination this would be A4 size.

More recently, satellite photographs have been used. They have proved most popular in meteorological questions where they can be used in conjunction with the weather map (or synoptic chart).

# 3

# POPULATION GEOGRAPHY

## POPULATION DEVELOPMENT

## DEMOGRAPHY

## GETTING STARTED

Geographers study the spatial variations in population in terms of the **demographic characteristics** of a population, i.e. the **structure** of a population in terms of its age, sex, marital status, family and household size. The ethnic composition, rates of growth and socio-economic characteristics may also be studied. A population is dynamic: it it changing constantly; it is the flow of people through time.

Population movement or **migration** occurs at a variety of scales – **intercontinental, inter-regional, intra-urban** – and for varying time periods – **permanent, semi-permanent** (e.g. shifting cultivators) or **ephemeral** (e.g. the pilgrimage to Makkah for the Hadj among Muslims, or even commuting as a diurnal migration).

This section is optional for those studying WJEC, Oxford and Cambridge, and London 16–19, so if you are not studying it please go to the following chapter (though you may still find this chapter of interest). For some of you studying JMB 'C', population is studied in the context of your regions and from a limited viewpoint. However, for those on all other syllabuses, population geography and demography is often one of the syllabus sections that you may study early on. Careful revision of material from Year 12 (Lower VI) is essential to make sure that it is at the right standard. If you are studying for the Scottish Higher there is a special section which relates population problems to health, disease, poverty, illiteracy and deprivation. Some of these issues are covered in other sections.

# ESSENTIAL PRINCIPLES

POPULATION
DEVELOPMENT

" Be familiar with this
model. "

The **demographic transition** or **population development model** (Fig. 3.1) suggests how the population of a country will develop through time and how these processes are linked to development (chapter 12).

This model is a broad generalization (like all models) and is only an approximation to the reality experienced by many countries in the Developed World. However, we cannot prove that countries in Stages I or II will follow the same path. You should know the characteristics of each stage and be able to give examples of countries at each stage.

Fig. 3.1 The stages of the
demographic transition

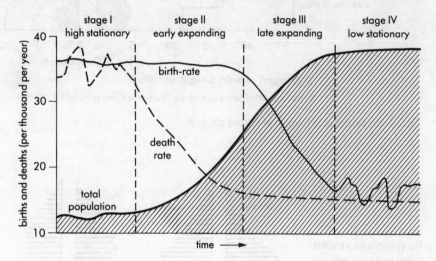

## POPULATION EXPLOSION

The **population explosion** is a major concern for the whole world: it is a rapid expansion of population in an area caused by a sudden decrease in the **death rate** but made more dramatic by a simultaneous increase in the **birth rate**. It is associated particularly with Stage II of the demographic transition. The possible effects of a population explosion have concerned people since Thomas R. Malthus wrote his first *Essay on the Principle of Population* in 1798, and more recently in the report of the Club of Rome called *The Limits to Growth* (1972).

The population problem is basically whether the resources of a country are able to support the population living there. This interaction is the subject of Chapter 4.

DEMOGRAPHY

Demography is the name given to the science of the study of populations. It focuses upon certain basic statistical data on populations normally gathered from censuses which are held in most countries every decade. Other data sources are used, such as surveys carried out by official bodies. Increasingly in countries which have decided *not* to hold censuses, such as West Germany and Denmark, registration data provide important information. The most basic data of demography is the number of people in an area at the time of the census (1991 in Great Britain). However, unless we organise that data, by itself it tells us little.

## AGE–SEX PYRAMID

" The age–sex pyramid
gives useful insights. "

We begin to understand the structure of that population by the age–sex pyramid which shows the make-up of the population by sex and age group. Fig. 3.2 shows such a pyramid for the city of Cologne in 1980, together with reasons for the particular changes in the shape of the pyramid. Are you familiar with the shape of the population pyramid for (1) a developed country such as Great Britain, (2) less developed countries, (3) smaller areas in cities, (4) areas experiencing migration, all at different stages in time?

Fig. 3.3 shows how the population pyramid in Great Britain has changed. Can you relate these changes to the stages in the demographic transition model? The population pyramid is a key way of beginning to analyse a population because it gives us clues as to how the population has changed, and is changing, through time.

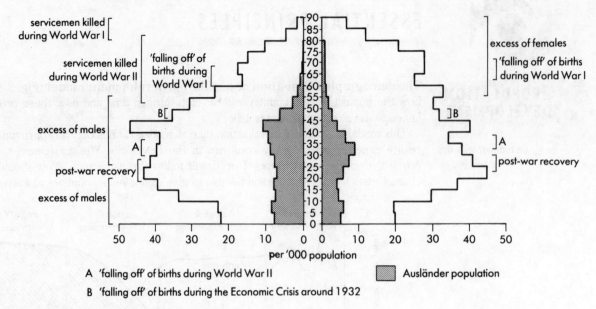

A  'falling off' of births during World War II
B  'falling off' of births during the Economic Crisis around 1932

Ausländer population

Fig. 3.2  The population of Cologne, by age and sex, 1980

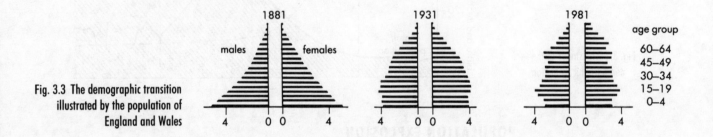

Fig. 3.3 The demographic transition
illustrated by the population of
England and Wales

1881        1931        1981

males    females

age group
60–64
45–49
30–34
15–19
0–4

4    0  0    4      4    0  0    4      4    0  0    4

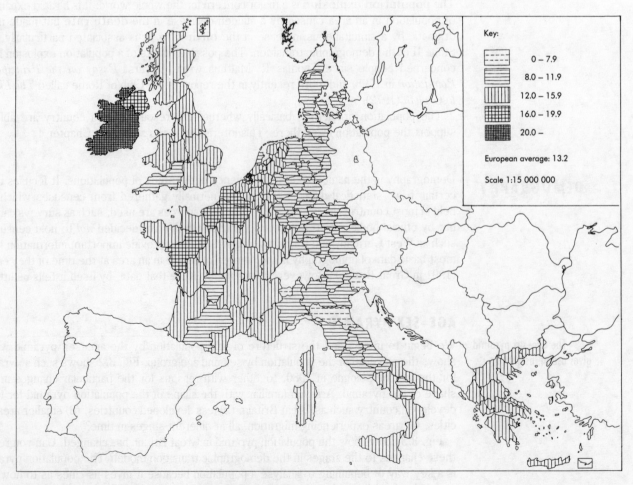

Key:

0 – 7.9
8.0 – 11.9
12.0 – 15.9
16.0 – 19.9
20.0 –

European average: 13.2

Scale 1:15 000 000

Fig. 3.4  Birth rates in the EEC, 1981 (*Source*: Eurogeo 2 1985, Eurogeo Working Group)

## BIRTH RATE

The **crude birth rate** (Fig. 3.4) is the number of babies born per year per thousand of population. It is an index of **fertility** or the ability of a population to reproduce. The rate is called 'crude' because it includes all people, including men, as well as old and young women who are unlikely to give birth. It does not allow for the different chances of giving birth. Therefore, the **age-specific birth rate** can be calculated either for each five-year age group or for the women of childbearing age. The birth rate is influenced by a number of factors:

1  The number of women of childbearing age, normally 15–45;
2  The age of marriage;
3  The number of married women;
4  The norms within a society for family size;
5  The availability and acceptance of methods of family planning;
6  The levels of education of society and especially of women in that society;
7  Economic conditions in the area.

Fertility varies with age, the number of existing children, the occupation of the mother, the type of area (e.g. rural or urban), medical knowledge, fashions in family size or the necessity to have children of one particular sex for social reasons.

## DEATH RATE

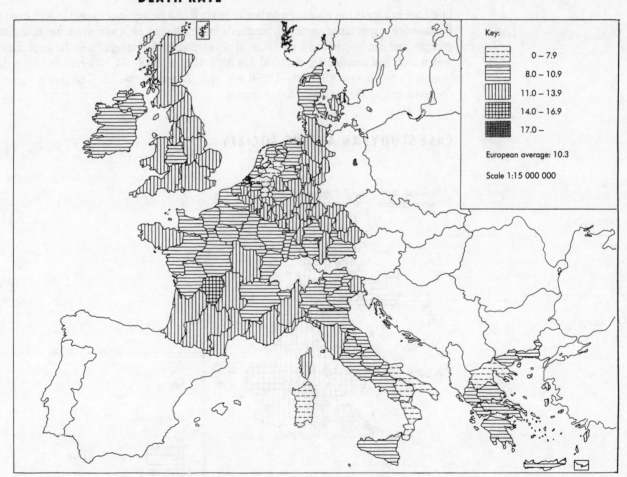

Key:

| | |
|---|---|
| - - - - | 0 – 7.9 |
| ≡≡≡ | 8.0 – 10.9 |
| ‖‖‖ | 11.0 – 13.9 |
| ⊞⊞⊞ | 14.0 – 16.9 |
| ▦▦▦ | 17.0 – |

European average: 10.3

Scale 1:15 000 000

Fig. 3.5 Death rates in the EEC, 1981 (*Source* as Fig. 3.4)

The **crude death rate** or **mortality rate** (Fig. 3.5) is the number of deaths per year per thousand population. It does not consider the age or sex of the population. The rate is of course influenced by these two variables, and also by social class, marital status, occupation and life-style. **Infant mortality** is an **age-specific death rate** measuring the number of deaths to children under one year old per thousand live births. It is an important indicator of health care in a country or area and therefore tells us about the degree of development of a society (see chapter 12). **Neo-natal mortality** refers to deaths in the first four weeks of life. Reductions in the latter rates are the result of medical advances,

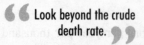 Look beyond the crude death rate.

improved nutrition of mother and baby and the general eradication of diseases by medical science. **Peri-natal mortality** is death either of an unborn baby (foetus) or during labour and birth.

## NATURAL INCREASE OF POPULATION

The **natural increase** of a population is the excess of births over deaths. **Natural decrease** is the opposite. The rate of natural increase is often shown as the gap between the birth and death rates on the demographic transition model (see Fig. 3.1) and will indicate the stage on the model reached by an individual country.

The worldwide natural increase in population is estimated at 85 million people per year (the equivalent of the population of Pakistan or Nigeria). In Nigeria the crude birth rate is 29/1000 and the crude death rate 11/1000 so that natural increase is 18/1000 per annum. In West Germany the crude birth rate in 1980 was 10/1000 and the crude death rate 12/1000 so that there was a natural decrease of 2/1000, giving an ageing population. Even a 3/1000 natural increase can be a problem, because for Egypt that represents 1 305 000 extra mouths to feed each year; a 2/1000 increase in Indonesia results in 2 976 000 extra people to be supported.

## LIFE EXPECTANCY

Life Expectancy is the average number of years that a person can expect to live from birth. Female life expectancy generally exceeds male. It can be expressed for different age groups, and can be used as a measure of development. In Bangladesh, the expectancy for males is 47 and females 48; in Nepal it is male 45 and female 44; whereas in Switzerland it is male 74 and female 78. In the USSR it is male 70 and female 75. Changes in these rates do result in specific problems for countries.

## CASE STUDY: AN AGEING SOCIETY

Ireland: data not available

Fig. 3.6 Age structure of the UK population 1982 (Population of 65 years or over as % of total population) (*Source* as Fig. 3.4)

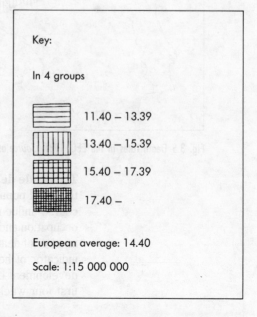

Key:

In 4 groups

| | |
|---|---|
| | 11.40 – 13.39 |
| | 13.40 – 15.39 |
| | 15.40 – 17.39 |
| | 17.40 – |

European average: 14.40

Scale: 1:15 000 000

> **This is an issue of current importance.**

In 1985, 15.1 per cent of the UK population were aged over 65; by 2020 that figure will be 18.7 per cent, and by 2050 projections suggest that the figure will be as high as 29 per cent. The distribution of the pensionable aged population in 1981 is shown in Fig. 3.6, and it is obviously focused in the South-West (where it grew by 13 per cent in the decade 1971–81) and in other coastal and rural areas. These concentrations are the result of **retirement migration** to the south coast in particular. In other areas some in-migration together with out-migration of the young has left high numbers of elderly, e.g. Wales. There is also a large number who have retired and remained in the same settlement because of low income; this is characteristic of the inner areas of many large cities. The effects of concentrations of the elderly have forced towns like Bournemouth to restrict the conversion of properties into old people's homes or to diversify their economies in order to attract a younger element – Bournemouth attracted Chase Manhattan Bank. Nevertheless there are firms who have prospered in the town by providing housing geared to the needs of the elderly. Health facilities will progressively be more stretched and retailing will have to adjust: for instance, old people buy fewer consumer durables.

The converse of the rise in the elderly is the fall in the number of young people entering the job market. Therefore some companies have shown themselves willing to employ pensioners in order to fill jobs formerly taken by teenagers. There are many other implications of the changing age structure of the UK which you could note, even in your home area. This will be an issue of concern, especially over the next decade.

## DEPENDENCY RATIO

The dependency ratio is a measure of the proportion of a population which is potentially active and employed. It is the *sum* of children under 14 and those aged over 60 (or 65) *divided by* the number of adults aged 15–64. A high ratio can indicate either a country with a rapidly expanding population at Stage II or a country in Stage IV (or even Stage V: declining population) of the demographic transition model. Fig. 3.7 maps old age dependency in 1982 for the European Community.

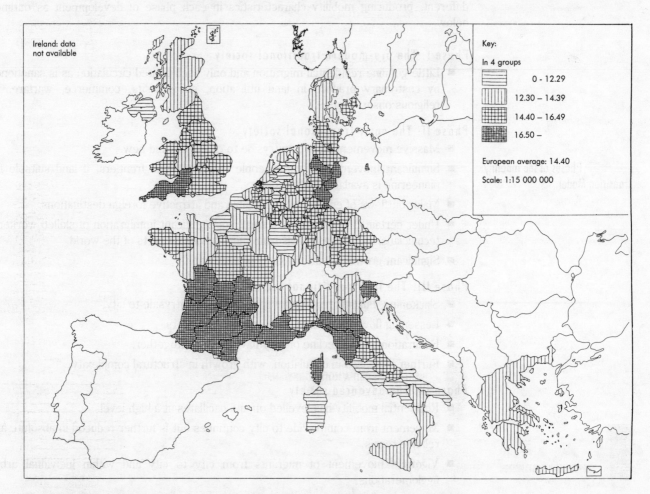

Key:

In 4 groups

| | |
|---|---|
| ☰ | 0 – 12.29 |
| ▥ | 12.30 – 14.39 |
| ▦ | 14.40 – 16.49 |
| ▨ | 16.50 – |

European average: 14.40

Scale: 1:15 000 000

Ireland: data not available

Fig. 3.7 Population of 65 years or over in the EEC, 1982 (as % of total population) (*Source* as Fig. 3.4)

**Remember the various definitions of labour force.** The **active population** is a term commonly used to refer to all the people of working age, normally 15–64, although some prefer the stricter definition of the **working population** (those in work and those temporarily unemployed) as a better measure of the **labour force**. The **activity rate** is the proportion of a population (total, male or female) which is gainfully employed. (Some might query the use of the term 'gainfully employed' because work in the home, usually performed by women, is essential to the functioning of the economy, releasing a partner for employment.)

## MIGRATION

**Migration is a key element in population change.** Population change is not solely the result of the interplay of births and deaths, natural increase and decrease. **Migration** is probably the most significant factor in population change. It includes a variety of flows of people, such as the mass movement of people to the New World in the nineteenth century. It also includes the movement of people seeking work or even a change of residence by moving to the suburbs, as well as the movement of the shifting cultivator in developing countries. **Mobility** involves travelling distances from a home base, whereas **migration** normally refers to a permanent or semi-permanent change of residence.

**Immigration** will cause a population to grow, both as a result of the new people themselves and because immigrant populations are generally younger and more fertile. **Emigration** may result in population decline, although only rarely does it cause absolute decline (e.g. Ireland during the nineteenth century.)

Within a country or region, migration alters the population distribution and the age–sex structure of the receiving and exporting regions. In the UK, migration into New Towns has generally been among young adults who then have children, whereas migration to the South-West has been characterised by retired persons.

## MOBILITY TRANSITION MODEL

Mobility transition is a five-stage model devised by Zelinsky which links to phases of development (chapter 13). In each category of migration the pattern of mobility is different, producing mobility characteristics in each phase of development as outlined below.

### Phase I: The pre-modern traditional society

- Little genuine residential migration and only such limited circulation as is sanctioned by customary practice in land utilisation, social visits, commerce, warfare, or religious observances.

### Phase II: The early transitional society

- Massive movement from countryside to cities, old and new.
- Significant movement of rural people to colonisation frontiers, if land suitable for pioneering is available within country.
- Major outflows of emigrants to available and attractive foreign destinations.
- Under certain circumstances, a small but significant immigration of skilled workers, technicians, and professionals from more advanced parts of the world.
- Significant growth in various kinds of circulation.

### Phase III: The late transitional society

- Slackening, but still major, movement from countryside to city.
- Lessening flow of migrants to colonisation frontiers.
- Emigration on the decline or may have ceased altogether.
- Further increases in circulation, with growth in structural complexity.

**Phases of the mobility transition Model.**

### Phase IV: The advanced society

- Residential mobility has levelled off and oscillates at a high level.
- Movement from countryside to city continues but is further reduced in absolute and relative terms.
- Vigorous movement of migrants from city to city and within individual urban agglomerations.
- If a settlement frontier has persisted, it is now stagnant or actually retreating.

- Significant net immigration of unskilled and semi-skilled workers from relatively underdeveloped lands.
- Vigorous accelerating circulation, particularly the economic and pleasure-orientated, but other varieties as well.

### Phase V: A future super-advanced society

- There may be a decline in level of residential migration and a deceleration in some forms of circulation as better communication and delivery systems are instituted.
- Nearly all residential migration may be of the inter-urban and intra-urban variety.
- Some further immigration of relatively unskilled labour from less developed areas is possible.
- Further acceleration in some current forms of circulation and perhaps the inception of new forms.
- Strict political control of internal as well as international movements may be imposed.

## RAVENSTEIN'S LAWS OF MIGRATION

These were developed on the basis of migration for Great Britain between 1871 and 1881. He outlined eleven laws as follows:

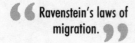
Ravenstein's laws of migration.

- The majority of migrants go only a short distance.
- Migration proceeds step by step.
- Migrants going long distances generally go by preference to one of the great centres of commerce or industry.
- Each current of migration produces a compensating counter-current.
- The natives of towns are less migratory than those of rural areas.
- Females are more migratory than males within the kingdom of their birth, but males more frequently venture beyond.
- Most migrants are adults; families rarely migrate out of their country of birth.
- Large towns grow more by migration than by natural increase.
- Migration increases in volume as industries and commerce develop and transport improves.
- The major direction of migration is from the agricultural areas to the centres of industry and commerce.
- The major causes of migration are economic.

Are these laws still true today?

## LEE'S LAWS OF MIGRATION

These distinguish four groups of factors underlying the decision to migrate:

1  Factors linked to the destination of the migrants;
2  Factors associated with the area of origins of the migrants;
3  The intervening obstacles between origin and destination;
4  Personal factors.

Do you know examples of these four? Some movement is related to *distance* – the greater the distance the less the numbers involved. Movement can be related to Zipf's **model of spatial interaction** and the **gravity model** derived from Newton's law of universal gravitation.

$$M_{ij} = \frac{P_i + P_j}{D_{ij}} \quad \text{where } M_{ij} = \text{movement between two places (i and j)}$$

$$P_i + P_j = \text{population of the two places}$$
$$D_{ij} = \text{distance between the two places}$$

## STOUFFER'S INTERVENING OPPORTUNITIES

Stouffer suggested a modification of the intervening obstacles. The number of migrants was directly related to the number of opportunities and inversely related to the number of intervening obstacles. The opportunities were factors such as housing, jobs and the environment.

## THE PUSH–PULL MODEL OF MIGRATION

This has been developed as one simple way of explaining many types of migration. Carr has merged his model with Lee and Stouffer to show the factors affecting the decision to migrate (Fig. 3.8).

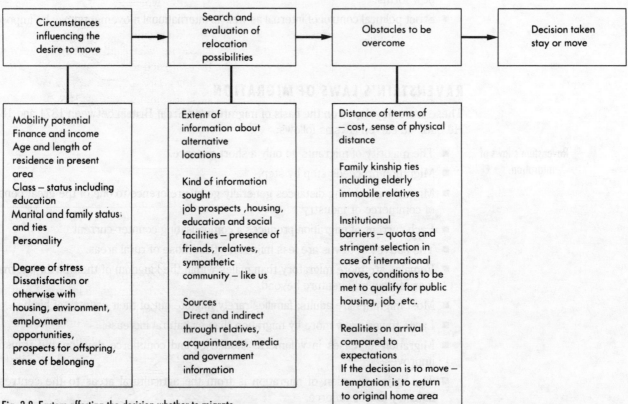

Fig. 3.8 Factors affecting the decision whether to migrate

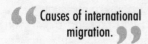 Causes of international migration.

**International migration** is caused by:

- Population pressure and inadequate resources (e.g. southern Italy to the USA in the early twentieth century).
- Shared poverty (e.g. 1840s Ireland).
- Better employment prospects (e.g. Turks moving to West Germany in the 1980s).
- Political factors (e.g. East German flight to West Germany, 1989).
- Religious persecution (e.g. Huguenots to England from France in the seventeenth century).
- Racial prejudice (e.g. the Ugandan Asians to Great Britain in the 1970s).
- Better value placed on skills (e.g. the brain drain from Britain to the USA).
- Ties between countries deriving from colonial times (e.g. West Indies and the UK, and Algeria and France).
- Freedom of labour movement (e.g. within the European Community).

## CASE STUDY: LABOUR MIGRATION TO WEST GERMANY

The booming West German economy of the 1960s could not rely on refugees fleeing from East Germany after the building of the Berlin Wall and so it turned to the Mediterranean countries to attract labour. By 1973, when West Germany decided to control the flow, 6.7

per cent of the population was foreign born and comprised 9 per cent of the workforce. In 1985 the foreign population came from Turkey (1.5 million), Yugoslavia (0.61 million), Italy (0.56 million), Greece (0.29 million), Spain (0.16 million), with a further 0.24 million from Asia and 0.38 million from Austria, Netherlands and Portugal combined. In some cities well over 10 per cent of the population are foreign and in some districts, especially of cities in the Ruhr (like Duisburg), there are areas inhabited almost exclusively by Turks living in company housing. Because the immigrants are young adults, many have young families: 22 per cent of the total foreign population is under 15 years old, compared with 15 per cent of the German population. Between 10 and 13 per cent of all births are to foreigners.

The foreign workforce is often concentrated in particular jobs where there is an element of danger or risk to health, such as iron and metallurgy (17 per cent of workforce) and plastics, rubber and asbestos (18 per cent). There are also major concentrations in the hotel and restaurant trades which are mainly comprised of Italians and Yugoslavs. The workers concentrate in the big cities, such as Frankfurt where 24 per cent are foreigners, with Italians dominating in Stuttgart, Portuguese in Hannover and Spaniards in Darmstadt. Turks are strong everywhere but especially in the Ruhr cities and in any city with major manufacturing plants, such as Stuttgart and Cologne where every third foreigner is Turkish. Within each city there are concentrations of each foreign population in different parts of the city. In Cologne, the nineteenth century city and the industrial suburbs of Mulheim and Ehrenfeld have large concentrations, although urban rehabilitation is forcing the Turks out of the inner city and into peripheral housing estates.

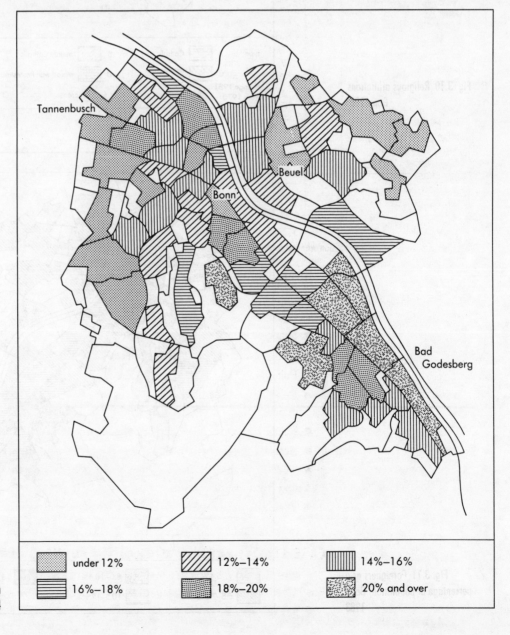

Fig. 3.9 Percentage of population aged 65 years or over, Bonn 1984

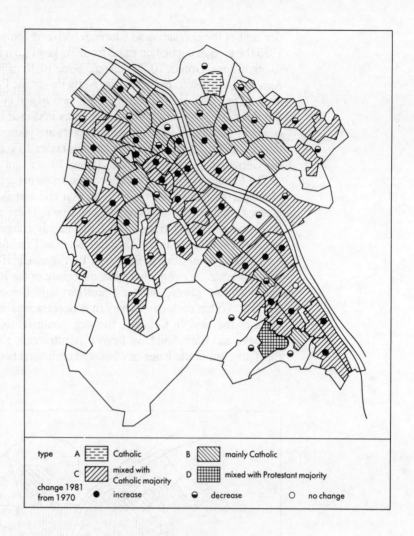

Fig. 3.10 Religious affiliations,
Bonn 1981

| type | A | Catholic | B | mainly Catholic |
|---|---|---|---|---|
| | C | mixed with Catholic majority | D | mixed with Protestant majority |

change 1981 from 1970   ● increase   ◒ decrease   ○ no change

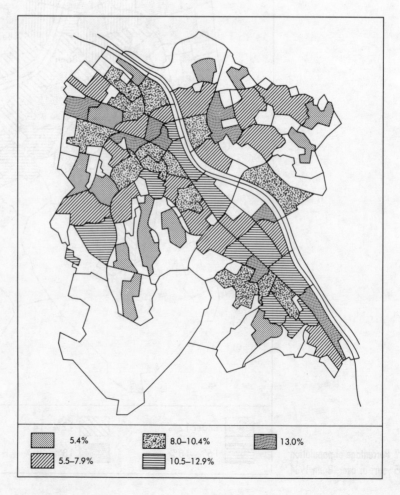

Fig. 3.11 Foreigners as a
percentage of population, Bonn
1983

| | 5.4% | | 8.0–10.4% | | 13.0% |
|---|---|---|---|---|---|
| | 5.5–7.9% | | 10.5–12.9% | | |

> ❝ Different forms of internal migration. ❞

**Internal migration** is characterised in a number of ways:

1  Rural to urban migration – i.e. **urbanisation** associated with the Industrial Revolution or with twentieth-century urbanisation without industrialisation in the Developing World (chapters 11 and 12). **Rural depopulation** resulted from the loss of agricultural employment, and it has often left behind a region of low population density, with declining services and infrastructure.

2  Urban to rural migration – this takes place in the area around major cities, giving rise to **commuter villages** and, since the 1970s, to the phenomenon of **counter-urbanisation**. Counter-urbanisation is the growth of an urban population and urban employment in small towns and villages at increasing distances from the major cities.

3  Intra-urban migration – the movement of population to the suburbs. **Suburbanisation** was often encouraged by transport improvements such as the electrified commuter lines in south London in the 1930s.

4  Urban to urban migration – for instance, the drift of young adults from the declining industrial cities, such as Liverpool, to London.

Check that you know some *causes* and *consequences* of these different types of internal migration. What are the effects on the population pyramids of both the areas of loss and the areas of gain?

Other characteristics of populations can be mapped if the data is available, especially by using the small area statistics which are available for the census Enumeration Districts (EDs). Not all data in the census is based on every household; some is based on a 10 per cent sample. Nevertheless it is possible to map the proportion of socio-economic groups, immigrants, migrants and religious groups. Figs 3.9–3.11 show some such patterns for the city of Bonn.

## PROJECTS

The theme of population is not an easy one to develop for projects. It is more probable that census data is used as a part of a study, e.g. of a commuter village. It might be possible to examine the distribution of an immigrant group from the census or from electoral rolls. You could then look for **field evidence** of their presence in terms of retail facilities, specialist services, or cultural facilities such as clubs and religious meeting places. Such a topic is very sensitive and needs handling with care and tact.

# EXAMINATION QUESTIONS

### Question 1
*Either*

a) To what extent do migrations between countries, and between regions within countries, have similar causes and consequences? Illustrate your answer by reference to specific examples. *(25)*

*or*

b) Table 3.1 shows average annual population growth rates by type of district in England and Wales, 1977–81 and 1981–86.

    a) Using the graph paper provided, draw a scatter graph that shows the growth rates in the two periods. *(6)*

    b) By reference to the graph, suggest and justify a classification of the types of district based on the growth rates in the two periods. *(6)*

    c) Discuss the factors which might explain the changing growth rates in the classes you have just suggested. *(13)*

(Oxford 1989)

| TYPES OF DISTRICT | GROWTH RATE PER THOUSAND POPULATION | | POPULATION AT MID- |
| --- | --- | --- | --- |
| | 1977–81 | 1981–86 | 1986 |
| England and Wales | 1 | 2 | 50 075 |
| Greater London | –7 | –1 | 6 775 |
| Inner London | –15 | –3 | 2 512 |
| Outer London | –3 | 0 | 4 263 |
| Metropolitan counties | –4 | –3 | 11 166 |
| Principal cities | –8 | –5 | 3 465 |
| Other metropolitan districts | –3 | –3 | 7 701 |
| Non-metropolitan counties | 5 | 4 | 32 134 |
| Large cities | –4 | –4 | 2 781 |
| Smaller cities | –1 | –3 | 1 718 |
| Industrial areas: | | | |
| Wales and the North | 0 | –3 | 3 327 |
| of which S. Wales valleys | –6 | –10 | 201 |
| Remainder of England | 4 | 2 | 3 370 |
| New Towns | 14 | 8 | 2 281 |
| Resort, port, retirement | 4 | 9 | 3 522 |
| Urban mixed urban/rural | | | |
| outside South-East | 7 | 8 | 3 990 |
| In South-East | 8 | 5 | 5 900 |
| Remote, mainly rural | 7 | 8 | 5 245 |

Table 3.1 Average annual population growth rates by type of district 1977–86, England and Wales

## Question 2

Study the population graphs and pyramids for Mexico and Japan (Fig. 3.12).

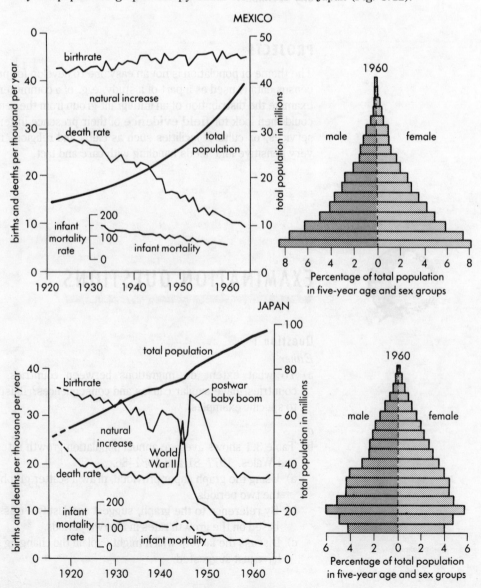

Fig. 3.12

a) What stage had each country reached by 1960 in the demographic transition (population development) model? (2)

b) Outline THREE major demographic differences between the two countries. (3)

c) With reference to Japan, how do you explain the apparent contradiction of a declining birth rate and an increasing population? (3)

d) Suggest THREE factors which might possibly contribute to the falling death rate in Mexico. (3)

e) i) For Japan, describe the characteristics of the pyramid for the age group under 30 years. (3)

ii) Account for these characteristics. (4)

f) i) In the space provided, Fig. 3.13, sketch a population pyramid for a country with an ageing population. (3)

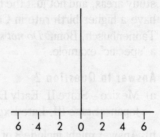

**Fig. 3.13**

ii) For a *named country*, give THREE reasons for such a population structure. (4)

(London, AS Specimen)

## Question 3

a) Study Fig. 3.14 which shows changes in birth rates and death rates in developing countries between 1900 and 1980.

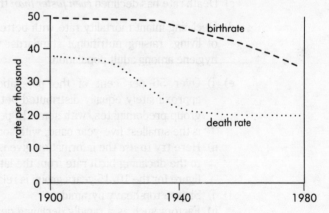

**Fig. 3.14**

i) How would you account for the changes in birth rates and death rates?

ii) How would these changes affect population growth? (8)

b) Describe the problems and the opportunities presented by rapid population growth in developing countries. (12)

(London, AS Specimen)

## Question 4

Discuss the contention that most migration can be explained in terms of a move to better opportunities.

# OUTLINE ANSWERS

### Answer to Question 1a

The question asks 'to what extent', so you must be evaluative. It is about the explanations of international and inter-regional migration, so show that you know the difference. Lee's laws and the concept of intervening opportunity can be applied at all scales. Many of the push and pull factors are common, although their relative importance alters. Population pressure affects both international and regional, but the former more. Shared poverty affects both, e.g. nineteenth-century Ireland and emigration, and migration from Sahel to Khartoum. Employment pulled people to the USA and pulls people south in the UK. Persecution has a greater international impact, although prejudice has pushed many Turks to the outskirts of German cities where they are not seen. Colonial ties are mainly a force in international migration, as is freedom of labour movement. It is also possible to note that Ravenstein distinguishes short- and long-distance migration, although his theories are less applicable today.

The consequences are loss of the young, the enterprising and (perhaps) one sex (this varies between short and long distance), the ageing of the society left behind, the focusing upon immigrant areas (is more characteristic of international moves) and the more rapid increase of population in the destination areas. All of this needs *examples* from your own study areas, and not just the bare principles outlined here. For instance, say that the Turks have a higher birth rate in Cologne than the Germans and show how they concentrate in Tannenbusch, Bonn. *Do not say* 'as in Germany'; that is awareness but does not constitute a 'specific' example.

### Answer to Question 2

a)  Mexico – Stage II, Early Expanding.
    Japan – Stage III, Late Expanding.

b)  Answer might include 3 of the following:
    - Mexico – rising stable birth rate; Japan declining birth rate. (Japan experienced post-war baby boom in late 1940s – not so in Mexico).
    - Both have a declining infant mortality rate, but Japan's has declined furthest.
    - Similarly both have a declining death rate, but Japan's has declined furthest.
    - Mexico has a much younger population structure than Japan, etc.

c)  Death rate has declined *even faster than* the birth rate over much of the post-war period.

d)  Declining infant mortality rate with better pre- and post-natal care; increased standard of living, raising nutritional standards; better general health care and standards of hygiene among adult population.

e)  i)  Over 56 per cent of the population are in the age group under 30 years, approximately equally distributed between male and female. The 10–15 years age group predominates, with some 12 per cent of the population. The 0–5 years group is the smallest five-year band, with some 8 per cent of the population, etc.
    ii) Here try to *use* the information given. Relate the low figure for the 0–5 years group to the declining birth rate from the late 1940s. The pyramid is for 1960, so the high figure for the 10–15 years group is related to the post-war baby-boom, etc.

f)  i)  Show a top-heavy pyramid.
    ii) Factors such as a rapidly declining death rate can be mentioned. This can be linked to improved health care, hygiene awareness, standard of living, etc. A declining birth rate in more recent times could be a factor in an ageing population, reducing the proportion of younger people. Birth control, higher standards of living, etc., might be mentioned. Three reasons linked to declining death/birth rate should be mentioned.

## STUDENT ANSWER WITH EXAMINER COMMENTS

### Question 4

> The statement "Most migration can be explained in terms of a move towards better opportunities" can be looked at upon 2 levels. These are internal migration and international migration.
>   On an international scale, migration for better opportunities is quite a gamble with the persons leaving their home country to go to a foreign land for work. An example of this international migration is the Turkish workers going to West Germany. These workers were at first invited to go to W. Germany to work as W. Germans had had a

great loss of manpower due to the second World War. They are known as gastarbeiter meaning "guest worker". When the guest workers arrived in W. Germany hoping for a better life they found that although their ideas of moving for better opportunities were sound life wasn't going to be a bed of roses. At first everything went well and many found jobs and moved into mostly cheap housing in the inner city areas. The money they made from their jobs, however, was used to buy themselves luxuries like cars and while this was good for them they would have found themselves better off going back to Turkey and trying to stabalise the economy. As time went on more workers came in to Germany mostly males who had left their families behind later to call on them when they were settled. It is at about this time that things started changing for them. They found there was great resentment of the Turkish by the Germans who felt they were pinching their jobs. Also the second generation of Turks who now see themselves as German citizens are finding it difficult to get jobs now and many are going back to Turkey to face a similar situation.

So although the move for better opportunities is what swayed many Turks to leave their home country this plan has now backfired leaving many of the second generation to go back to Turkey. They have in fact not found any better opportunities after the first boom with there now being resentment, no jobs for them, cheap, slum-like housing in the inner city.

An example of internal migration again done in order for better opportunities is within France. In the Massif Central region in the S. East of France there were many agricultural problems such as an annual drought which reduced crops to a minimum and these problems had the affect of making many farmers sell up and think about moving into urban areas. They saw it as a move with better opportunities, better jobs, better housing, better education for second generation, more facilities and a better standard of living. However when they got to the city they found themselves with the same problems as the Turkish gastarbeiters. The city was overcrowded leading to no jobs especially for the unskilled, no housing only cheap slum-like housing, a lower standard of living rather than the higher standard they had left for and a more expensive life on the whole.

To combat this problem the French government has tried to make the South East areas where the people had left more attractive. They have built roads, making transport easier, they have enforced a standard dialect so everyone is equal with language. They have improved the educational facilities and many of the small unprofitable farms have been amalgamated in an effort to make them more profitable. As a result of these methods people are now moving back to these areas for the same reasons they left a "move towards better opportunities".

So although people migrate for better opportunities a lot find themselves worse off and then are stuck trying to get themselves out. It is rare to find situations in which whole communities find themselves in a much better situation to that which they left. This is perhaps the main fault with both the Turks and the mass exodus of French from the S. East.

**66** A reasonable illustration. **99**

**66** Can you say anything specific about the lack of opportunity in Turkey, as well as the openings in Germany? **99**

**66** Why not mention the term 'rural depopulation'? **99**

**66** Rather vague. Be more precise. **99**

**66** What about other factors affecting migration besides better opportunities? **99**

**66** This answer is on the right lines, but it does not demonstrate enough knowledge of theory, and the application of theory, to gain a high mark. **99**

CHAPTER

4

# POPULATION AND RESOURCES

## KEY CONCEPTS AND THEORIES

## THE POPULATION PROBLEM

## POPULATION POLICIES

## KEY CONCEPTS IN THE STUDY OF NATURAL RESOURCES

## GETTING STARTED

The relationship between population and resources is a key feature in human geography. The ability of a country to provide for itself through the utilisation of its natural resources, its industrial skills, and its knowledge and brain power, determines a country's ability to support a given population. Health standards and mortality indicate the relation between population and resources.

On the whole this section of your studies is more open to debate because there are contrasting opinions regarding policies: for instance, population control. Do not necessarily accept one particularly persuasive argument; instead, be prepared to understand alternative positions on some of the moral issues that are posed by the population problem. Those of you with good essay-writing skills and good knowledge could do well answering questions in this area of your syllabus.

'Global Limits to Growth' and 'Feeding the World's Population' are key themes in a number of syllabuses (e.g. London 16–19), although some of the material you need for these topics will be found in chapters 10, 12 and 13.

# ESSENTIAL PRINCIPLES

**KEY CONCEPTS AND THEORIES**

## MALTHUS'S THEORY OF POPULATION GROWTH, 1798

This has two principles:

1   Population grows at a *geometric* rate;
2   Food production increases at an *arithmetical* rate.

The consequence is that population growth will eventually exceed the capacity of agriculture to feed that growth. Population would rise until a ceiling to growth was reached when (1) *preventive checks* (e.g. delayed marriage) and (2) *positive checks* (e.g. famine, disease, war and infanticide) would increase the death rate.

There is evidence of Malthusian checks in the Developing World. However, in the Developed World, Malthusian predictions have been slow to be realised because:

> " Reasons why Malthusian theories are less relevant to the Developed World. "

a)  he did not anticipate the Industrial Revolution;
b)  he did not foresee the growth in agricultural output (especially in the New World and the tropics) and the development of new crops and new strains;
c)  the impact of transport and refrigeration came after his time;
d)  he confused moral and religious issues with population issues; and finally
e)  he could not foresee medical advances in birth control methods.

## BOSERUP'S THEORY, 1965

This suggested that 'necessity was the mother of invention', and that society only developed when it was threatened by a lack of resources. Therefore population pressure is necessary to ensure that people are forced to discover new agricultural methods and crop strains with increased yields.

## LIMITS TO GROWTH REPORT, 1972

This was prepared by the Club of Rome group and is *neo-Malthusian* in that it updates Malthus's ideas. Population growth is determined and limited by a number of factors:

a)  population;
b)  food production;
c)  natural resources;
d)  industrial production; and
e)  pollution.

The report concluded that if all of these five factors continued to grow exponentially then the capacity of the earth to sustain growth would be reached by 2070. At that time population and industrial capacity could both decline. They suggested that this conclusion would be avoided by policies of economic or ecological stability which would bring about a state of **global equilibrium**.

## POPULATION–RESOURCE RATIO

This is the relationship between the amount and quality of a country or area's natural resources and the size and competence of its population. It was developed by Ackerman who distinguished five types of **population resource regions**:

> " Types of population resource region. "

■ *The United States type* – advanced technological societies with low population growth and high availability of resources.

■ *The European type* – advanced technological societies with a high population–resource ratio.

■ *The Brazil type* – technologically deficient regions of low population–resource ratio.

■ *The Arctic–Desert type* – areas with poor technology and limited food-producing resources.

Much of this classification is dated because of the changing pattern of resources (e.g. oil being found in Alaska) and changing population growth rates (e.g. Western Europe's stable population).

## OPTIMUM POPULATION

This is the same as **optimum carrying capacity**, but a more ecologically based term; it is the ability of an area to support a population in such a way that the natural resources are fully utilised. Put another way, the per capita output is maximised within the prevailing technological and socio-economic constraints of the area. The optimum is of course *below* the survival level.

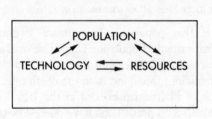

Fig. 4.1a) The basic relationship

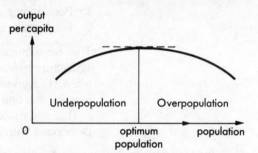

Fig. 4.1b) The relationship between population and output

**Overpopulation** occurs where the resources are unable to sustain a population at the existing living standard without a reduction in that population.

**Underpopulation** occurs where an increase in population would result in the more effective use of resources which would raise living standards.

All of these three concepts are difficult to measure because of the constantly changing resource base and the difficulties of measuring the resource value of certain industries, especially services.

## POPULATION DENSITY

Population density is the number of persons per square kilometre/hectare. This is influenced by both physical and human factors, but to varying degrees which depend on how advanced a particular society is.

> **Physical factors affect population density in developing countries.**

**Physical factors** are rarely important in explaining population density in developed countries, except in some rural areas (e.g. uplands). Physical factors may still influence the size of settlements, which depend on water, level sites, navigable rivers and mineral extraction for the development of economic activity. However, relief, climate, certain types of vegetation, wetlands and the presence or absence of mineral, food and energy resources, will all have effects on population density in a developing country. Do you have a case study of a developing country which illustrates the impact of physical factors on population density?

**Human factors** include the number of people to be supported, the age and health of the workforce and the education level. Human factors also influence the level of investment, the nature of the infrastructure and the political and religious characteristics, all of which might affect the density of population in different countries.

The way in which physical and human factors affect the density of population varies between developed and developing countries.

A third group of factors concern **location** and **access**: Is an area isolated or accessible? It is well integrated into modern global communications or remote from them?

> **Examples are important.**

For all of these factors you should have **examples** from both the Developed and Developing Worlds.

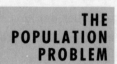

**THE POPULATION PROBLEM**

This is a social and moral issue as much as a geographical one. You will have your own opinions and so make sure that you can support them by reference to real examples and by using the terms, concepts and theories introduced in this chapter and the preceding one. This section comprises a set of maps which might help you to develop your ideas – little explanation is given here. The area of each country in Fig. 4.2 has been made proportional to its population. In Fig. 4.3 we can see the number of years projected for the doubling of population in the various countries.

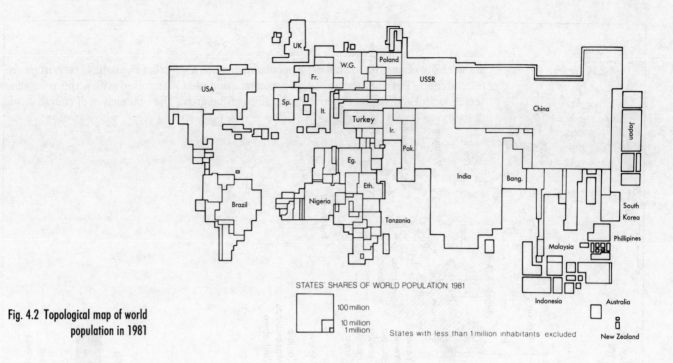

**Fig. 4.2 Topological map of world population in 1981**

STATES' SHARES OF WORLD POPULATION 1981

100 million
10 million
1 million

States with less than 1 million inhabitants excluded

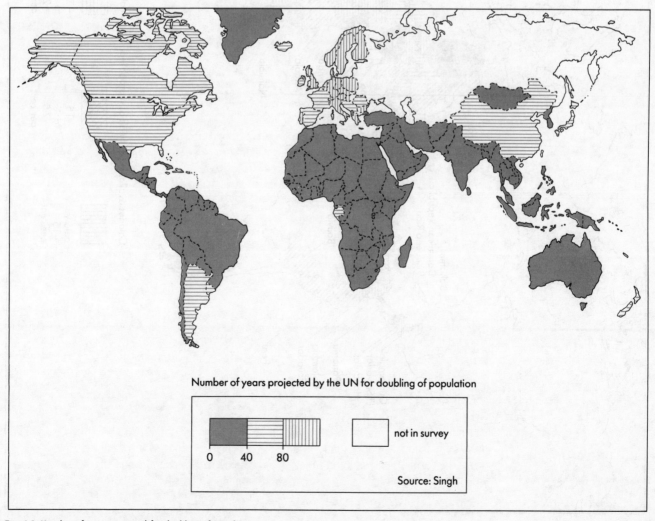

Number of years projected by the UN for doubling of population

0    40    80    not in survey

Source: Singh

**Fig. 4.3 Number of years projected for doubling of population**

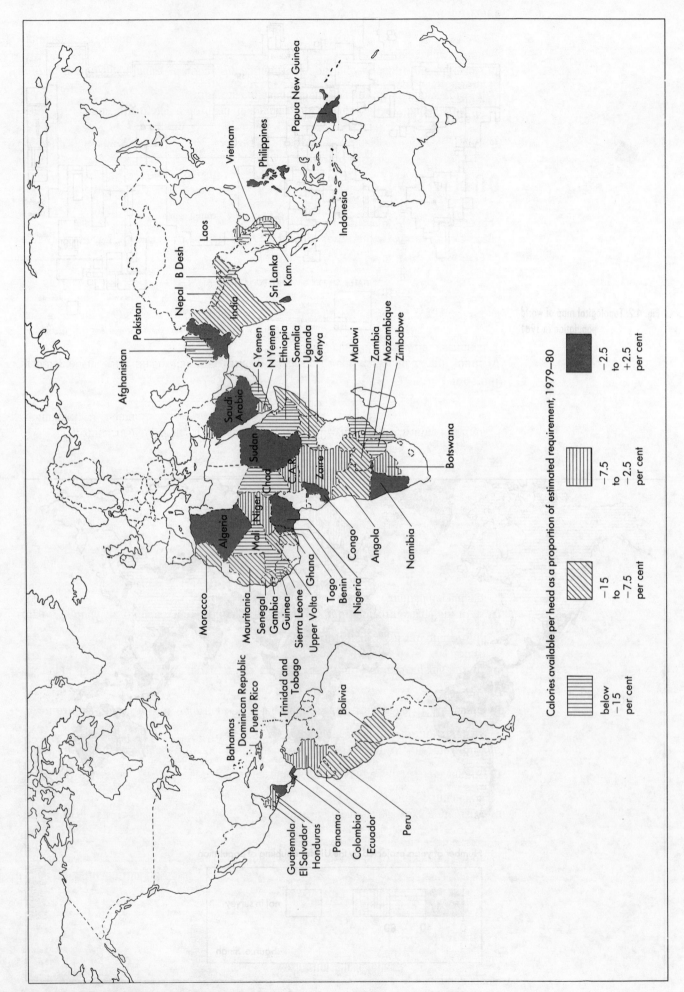

Fig. 4.4 Calories available per head as a proportion of estimated requirements, 1979–80

## CASE STUDIES: FEEDING THE WORLD

Related to the population problem is that of feeding the world of which, as Fig. 4.4 shows, approximately 60 per cent are underfed. Food resources are **overproduced** in some countries (e.g. the USA), while others cannot feed themselves (e.g. Chad). Countries overproduce because:

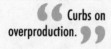
*Causes of overproduction.*

a) They want to gain export earnings but many cannot afford the price;
b) The government assists in marketing the crop;
c) The government subsidises the farmer with price supports which encourages overproduction;
d) The physical conditions suit the growing of a crop, e.g. the corn belt;
e) Agricultural technologies have enabled greater output, not least the use of oil-based fertilisers, mechanisation and plant and animal protection;
f) Irrigation has brought more land into use.
   All of this can result in:

1 Environmental degradation – soil erosion due to exposed fields;
2 Pollution;
3 The excessive use of energy resources, water and even soil.
   How is overproduction curbed or ameliorated?

*Curbs on overproduction.*

a) by government policies, such as the European Communities **set aside policy** to pay farmers *not* to grow crops,
b) by *halting* grants for marsh-draining, hedgerow removal, etc., which had been given in the interests of greater efficiency;
c) by using surplus as an aid commodity, so having a humanitarian policy which reflects well on the donor; and
d) by selling cheaply to other states (e.g. to the USSR or Poland), which again has political aims.

Countries cannot always feed themselves. The population of Lesotho is growing at between 15 and 33 per cent per decade, with the result that the majority are under 20 years old. Because the country is small (30 000 sq km), rugged (40 per cent mountainous), hilly (with only 25 per cent lowland) and climatically extreme (hot, wet summers, cold winters and a high risk of drought), the country hardly has the environment to support increased food supplies. Land is overused so that yields deteriorate, soil erosion accelerates and therefore people leave the land for the city of Maseru or to work in South Africa. Because all the land is owned by the king, the incentive to conserve it and to farm effectively is low. Even if the outdated land tenure system is overhauled the country will still have trouble feeding itself. How to improve such a country takes us into the realms of development, but is labour migration to South Africa the answer, with the returned earnings helping to pay for the imported food? Or are there other answers to the problem of feeding Lesotho?

## POPULATION POLICIES

These are the measures taken by governments to alter the rate of population change. They can be (1) **expansionist** or (2) **controlling**.

1 **Expansionist policies:**

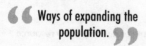
*Ways of expanding the population.*

- Restricting availability of birth control (e.g. Ireland, Ghana);
- Making abortion illegal;
- Giving family allowances and higher allowances for more children (e.g. France);
- Giving more support to working mothers (e.g. East Germany);
- Improving health care;
- Restricting roles for women in society (e.g. Fundamentalist Islamic States).

2 **Control policies:**

*Ways of curbing the population.*

- Introducing programmes of birth control;
- Allowing abortion;
- Encouraging sterilisation (e.g. India);
- Introducing government control over number of children (e.g. China);
- Reducing support for large families;
- Running government advertising and education campaigns (e.g. Singapore and India).

# PROJECTS

The theme of population and resources is probably too complex for you to attempt as an individual **field project**. However, it can be explored, particularly with reference to development, using **secondary** sources.

**KEY CONCEPTS IN THE STUDY OF NATURAL RESOURCES**

## STOCKS

These are the total components, undiscovered and discovered which make up the earth. Some of these, e.g. iron in the earth's core, and most energy resources, are unavailable to us.

## RESOURCES

These are items which can be of use to people for food, shelter or to gain a livelihood. **Reserves** are the amount of a resource available under the current technological and economic conditions. These reserves change as technology develops and as social and economic demands alter.

## NON-RENEWABLE RESOURCES

These are finite and will be exhausted. Coal, oil, natural gas and mineral ores are all non-renewable. Resources can be **non-recyclable**, e.g. coal, or **recyclable**, e.g. copper.

Renewable or flow resources are variable in output but always present, such as wind, tidal, solar or water energy. People have had minimal impact on these resources until recently. In contrast, human action can affect some flow resources such as forests and fish. If cutting and regeneration or replanting are not balanced then the timber resources may decline; but it is within human powers to rectify the balance or even to improve the total flow. Soil (chapter 9) is also seen as a renewable resource open to over-exploitation. Crops are a short-term renewable resource. Resources are defined by society's need, their economic viability and the technical capabilities available to exploit them.

## CYCLE OF EXPLOITATION

> **The cycle of exploitation.**

Most resources pass through a **cycle of exploitation**:

1  Exploration and discovery – reserves at their highest;
2  Exploitation of major finds – reserves increase;
3  Exploitation of minor finds – declining reserves;
4  Exhaustion – recycling and substitution.

Fig. 4.5 shows the cycle in theory.

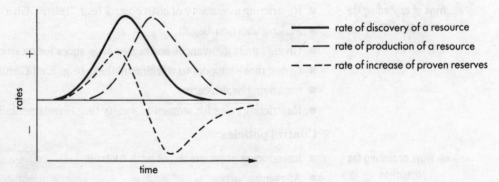

Fig. 4.5 Cycle of exploitation of a resource

—— rate of discovery of a resource
– – – rate of production of a resource
– ·· – rate of increase of proven reserves

rates

time

In the case of the North Sea gas and oil fields, stage (1) was triggered by the Middle East oil price hike of 1973 and the availability of offshore technology. Stage (2) has existed

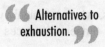

**Alternatives to exhaustion.**

from 1975 to the 1990s, when it is expected that stage (3) will be reached. The final stage will be reached in the first quarter of the twenty-first century.

When a resource is facing exhaustion there are several possibilities for the user:

- Discover alternative sources of the same resource (e.g. offshore oil-prospecting elsewhere around the UK);
- Import the resource;
- Conserve existing reserves by more efficient use of the existing resource (e.g. energy conservation measures);
- Where possible, recycle the material (e.g. lead, copper, aluminium);
- Find substitute materials (e.g. the aluminium can for the tin can, new types of battery, fibre optic cable).

## CASE STUDY: WORLD COPPER AND LEAD

|  | 1965 | 1970 | 1975 | 1980 | 1983 |
|---|---|---|---|---|---|
| World copper output | 5.04 | 6.03 | 7.0 | 7.7 | 8.0 |
| World copper consumption (Life expectancy of reserves: 20–56 years) | 6.11 | 7.26 | 7.47 | 9.36 | 9.11 |
| World lead output | 2.69 | 3.39 | 3.43 | 3.44 | 3.32 |
| World lead consumption (Life expectancy of reserves: 17–42 years) | 3.19 | 3.87 | 3.91 | 5.34 | 5.22 |

Table 4.1    *Source: World Resources 1986*

| COMMODITY | TIME IN YEARS UNTIL RESERVES EXHAUSTED |
|---|---|
| Bauxite | 51 |
| Chromium | 54 |
| Cobalt | 36 |
| Gold | 19 |
| Iron | 46 |
| Mercury | 14 |
| Nickel | 30 |
| Tin | 23 |
| Zinc | 15 |

Table 4.2

Table 4.1 shows that the output of both of these minerals falls short of consumption: in the case of lead by almost 2 million metric tonnes in 1983. The shortfall would have been worse for both minerals had output not increased in some countries to replace declining reserves among the traditional producers. For instance, although copper output in Japan, Zambia and the USA has declined, that from Chile and the USSR had more than replaced the decline. Less lead is now produced in Canada, Mexico and Sweden, but Australian output has more than replaced it. American lead output also rose in the 1970s, but is now declining. Increasingly the mineral content of the ores is lower and therefore more costly to extract. Recycling helps but, with diminishing reserves, substitutes will need to be found.

Other minerals have also been estimated to have limited reserves. Assuming a 5 per cent growth in demand, Table 4.2 shows how long it is estimated that some of the world's important mineral reserves will last. Despite these alarming figures, mineral availability in the medium-term is good. It is political and economic forces rather than physical stocks which often cause scarcity.

## RENEWABLE RESOURCES: THE CASE OF FORESTRY

People have always cleared the forests for agriculture: up to 50 per cent of the temperate forests have been cleared over the centuries, compared to a loss of about 6 per cent in the tropics. However, these trends mask recent changes in the extent of forests, which are growing in Europe and North America but are declining rapidly in Latin America and Africa. In tropical regions deforestation rates have *exceeded* reforestation rates by between 10 and 20 times. Up to 5 million hectares of the tropics are being felled for timber or for agricultural land each year. In Europe 92 per cent of forests are commercially productive, which reflects the long-term management of the forest resource. In contrast, 20 per cent of the productive *tropical* forest has been cut and only 50 per cent of the total can be regarded as being commercially exploitable.

If timber supplies are to be maintained, forest management is imperative to preserve these 'flow resources'. Despite the effects of acid rain (chapter 7) and pollution, 60 per cent of Europe's forests are managed according to a planned programme of felling and replanting. Only five African and four Asian countries have forest plans. In Africa only 1 per cent of the forest is managed. Even where forests are managed (as in Malaysia) and felling is regulated, the loggers often bribe the inspectors to ignore excessive felling. This reflects the rewards for felling more of the tropical hardwoods, which are highly valued in the Developed World. Yet without management, certain timber resources will be exhausted.

The policies to maintain forests are:

1   Replanting, either on a 1 : 1 basis or more new trees for every felled tree;
2   Reforesting the marginal land not needed for agriculture;
3   Improving the genetic quality of the trees – in Brazil the yield of eucalyptus has been raised by 150 per cent in 20 years; and
4   By methods of forestry that optimise the growth of the trees.

In some areas industrial and commercial tree crops replace forest, e.g. rubber and oil palm plantations in Malaysia, and so the ecological effects are reduced, even though timber revenue has been replaced by rubber and palm oil.

> Policies to maintain forests.

## PROJECTS

Projects in the field of resource management are almost impossible and best avoided. It might be possible to look at the topic by considering issues such as the use and re-use of the water resource, or by looking at the recreational resource value of heathland as opposed to open woodland and managed plantations.

# EXAMINATION QUESTIONS

### Question 1

In 1986 the International Institute of Environment and Development rated deforestation as one of the world's most pressing problems. Taking your examples from TWO global forest ecosystems, illustrate and explain the consequences of deforestation.

(AEB, November 1988)

### Question 2

With reference to specific areas, examine the problems of the use and management of water.                                                                           (AEB, June 1987)

### Question 3

Discuss the geographical significance of the distinction between renewable and non-renewable resources.                                                               (WJEC, June 1988)

### Question 4

Explain why the predictions contained in Malthus's theory of population have been slow to be realised.                                                                  (AEB, June 1989)

### Question 5

Study the following extract which is a precis of the strategies which might be taken to reduce the destruction of tropical forest.

a) What factors have initiated the forest destruction?                              (6)
b) Provide a table which identifies the advantages and disadvantages of the few strategies shown in the extract.                                             (12)
c) For an agricultural system in a *named* developed country identify the social and ecological impact of agricultural changes experienced in the last forty years.                                                              (7)

(AS London 16–19 Sample Paper)

## Strategies to end the destruction

The world possesses nearly two billion hectares of tropical forest. The 56 countries which contain the most seriously affected forests also hold about half the developing world's rapidly expanding population (*Richard North writes*).

Several of the biggest aid agencies and think tanks agreed two years ago that they should co-ordinate a Tropical Forestry Action Plan. This was refined this month when the bodies met in Italy.

The revised strategy stresses the needs of the 200 million forest people who live in tropical forests, and the need to encourage and listen to the emerging grassroots environmental movements in poor countries. The plan already involves a country-by-country analysis of the problems and opportunities facing the forestry sector in each one. Charles Secrett, forests campaigner for Friends of the Earth says: 'They still put far too much emphasis on the activities of the very poor in the forests, as though they were not there in part because of displacement by industrial agriculture elsewhere. There was no talk of the way the debt problem affects forests.'

There are now perceived to be several channels for forestry conservation:

■ Direct lending for conservation projects: support from the international community is likely to double to $1bn a year over the next year or so;

■ 'Conditionality': making development loans of every kind dependent on a country agreeing to put its forestry in order;

■ A new bureaucracy, the International Tropical Timber Trade Organisation, which might become the custodian of a sort of 'Green Forest Product' label system, to encourage a trade in timber grown sustainably. But Mr Secrett believes that we will need to help poor countries preserve forests, while sustainable forestry techniques are developed;

■ The possibility that ultimately the rich world may have to lease parts of forests from the country which owns it. Costa Rica and Bolivia are having some of their debt written off in exchange for establishing reserves.

*Source: The Independent*

CHAPTER **5**

# ENERGY

ENERGY USE

ENERGY RESOURCES

ENVIRONMENTAL
AND SPATIAL
CONSEQUENCES

ENERGY IN LESS
DEVELOPED
COUNTRIES

THE POLITICS OF
ENERGY

## GETTING STARTED

Energy use is normally a part of the resources or the economic geography section of most Geography syllabuses. The main sources of energy include both renewable (e.g. solar) and non-renewable (e.g. oil) resources, and so the topic is ideal for a case study in resource management. Similarly energy sources are a key factor in the location of industry (chapter 10) and some fuel-based industries, such as oil refining and chemicals, illustrate certain principles of industrial location.

The 'Energy Question' is a major theme in a number of syllabuses (e.g. London 16–19) and therefore it was felt necessary to distinguish energy from the rest of resources. In this case you need to know how people have obtained energy and with what environmental and spatial consequences. You also need to know how energy resources can be best managed in order to conserve them for the future and to reduce the effects of energy use on the environment.

**Energy** is the means of providing heat, light and power; it is the lifeblood of society because it drives almost all human activities. **Primary energy** includes the basic energy resources: coal, oil, solar power and water. All of these resources are converted to form **secondary energy**, of which electricity and petrol/diesel are the main products. These are then distributed by grid or pipeline to be used in heating, lighting, transport and for powering the machinery used for economic production. At all stages there is **energy loss**, in transformation heat, distribution leaks and in the efficiency of the machinery.

# ESSENTIAL PRINCIPLES

Energy flows through an economic system, beginning either as a natural resource or as an imported resource. It is focused and converted to raw materials for industry (e.g. plastics), agriculture (e.g. fertilisers) and transport (e.g. oil), or it is used to provide power, heat and light. These are used in producing finished products which use still more energy in being transported to the markets.

Energy use has varied through time. Before the Industrial Revolution, timber plus wind, water and animal power were important. But the invention of the steam engine brought about the dominance of coal as a source of heat and motive power, and subsequently the provision of coal gas for heat and light. From the late nineteenth century, the use of coal to generate electricity furthered its dominance in energy supplies. The development of the use of oil and its associated natural gas is a twentieth-century development. Atomic power is a product of the last forty years, although it does not as yet dominate energy supplies, except in the provision of electricity in countries lacking alternative sources (e.g. France).

Energy use also varies between countries. Fossil fuels – oil, gas and coal – account for 90 per cent of the global use of energy, and these are the major energy sources of the *industrialised* world. However, firewood, charcoal and biofuels (animal and crop residue) are the main energy resources for half of the world's population, or 2.5 billion people. More people depend on wood than on any other single source of energy, hence the importance of reforestation which we noted in chapter 4.

> Different energy sources have dominated at different times.

## Oil

Oil accounts for 40 per cent of commercial energy consumption. Its consumption in the Developed World is declining (from 70 per cent in 1973 to 57 per cent in 1984), in contrast to the Soviet bloc (14 per cent in 1973 to 19 per cent in 1984) and the Developing World (16 per cent in 1973 to 24 per cent in 1984). Most oil (54 per cent) is used in transport, with 20 per cent used in residences, 19 per cent in industry and the rest in the service sector and agriculture. There are only 12 major exporting countries, dominated by Saudi Arabia and other Middle Eastern states. In the 1980s the UK's oil trade was in balance. The imbalances in supply and demand for oil are a major political concern for the future. You need to know where oil is produced and consumed and what political factors influence the trade.

## Coal

Coal accounts for 30 per cent of commercial energy consumption and its contribution has risen because of the fears for oil supplies since 1973. Coal reserves are exceedingly large and will last over centuries; it has been estimated that it would take 1008 years to exhaust the supply at current rates of usage. Coal reserves are more evenly distributed throughout the world than oil reserves (USA 21 per cent, USSR 23 per cent, China 15 per cent). (See also Fig. 5.5.)

## Natural gas

Natural gas accounts for 20 per cent of energy consumption and its production is dominated by North America, Europe and the Soviet Union. Consumption is rising everywhere but especially in the USA, as natural gas is substituted for oil. The United Kingdom is self-sufficient, thanks to supplies from the North Sea. The reserves of gas are double those of oil, and with further discoveries in the Developing World they should last well into the next century. Where reserves are dwindling (e.g. the USA) gas from shale and coal-seam methane might be used as substitutes.

## Nuclear power

Nuclear power provides around 3.9 per cent of commercial energy, but over 90 per cent of nuclear power is generated in the Developed World. Nuclear power is unlikely to expand greatly because of (1) the high capital cost of building, (2) the high cost of decommissioning old stations, (3) the impact of disasters such as Chernobyl and Three Mile Island and (4) problems with the disposal of spent fuel.

> Problems for nuclear power.

## Hydro-electricity

Hydro-electric power (HEP) contributes 6.7 per cent of commercial energy, with 50 per cent coming from North America and Western Europe where the technology has been applied to most potential sites. The share in developing countries is also growing; it already contributes 41 per cent of electricity production in the developing countries. Capital costs and the costs of displacing people for the huge storage lakes (e.g. Aswan, Kariba, Akosombo) are deterrents to further development, except in sparsely peopled areas. Silting can be a problem, so reducing water storage (e.g. Aswan).

### Biomass energy

Biomass energy is mainly wood (6 per cent of energy consumption) but does include dung (90 per cent) in some Indian villages, and straw (HRH Prince Charles has a straw-burning system). Twenty-one countries depend on wood for 75 per cent or more of their energy needs and in Burkina Faso, Malawi and Mali, wood provides over 90 per cent.

> **Other energy sources.**

Other sources are relatively insignificant. **Solar power**, modern **wind generation** and **wave power** are technologies in their infancy which are only viable in the tropics, in areas of constant strong wind and in coastal regions, respectively. **Tidal barrages** (e.g. on the Rance in Brittany) are constrained by capital costs although there are many potential sites (e.g. the Severn estuary and the Bay of Fundy). **Geothermal power** is used at Larderello in Italy, the USA and the Philippines, and there is potential in New Zealand, Mexico and Kenya.

## ENVIRONMENTAL AND SPATIAL CONSEQUENCES

Conventional electricity power stations needed to be near the source of fuel, e.g. coal, and supplies of water to provide steam. Early stations were sited on rivers in the coalfields where both fuel and cooling water were available, and on the coast. Here coastal colliers could bring fuel, and sea water could be used in the turbines. As the economies of generation grew, so the sites have become more restricted.

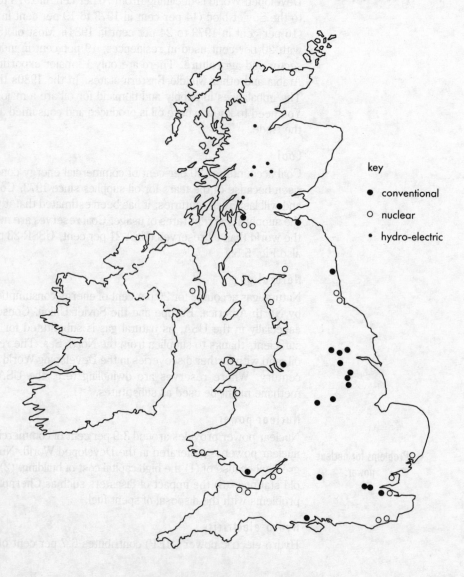

key

● conventional

○ nuclear

• hydro-electric

**Fig. 5.1 Power stations in the UK**

Water supply problems have meant the construction of large sets of vast cooling towers (as on the Trent at Long Eaton) with their plumes of cloud and inevitable micro-climatic effects. Large generators could best be transported by sea to coastal sites because many inland road bridges were not strong enough to take the weight of generators being transported to potential power-station sites. Cheap oil supplies in the 1960s led to the development of large oil-fired stations adjacent to refineries, e.g. Fawley. More recently, further proposals for oil-fired stations have been dropped, owing to the increased costs of fuel oil, and large coal-fired stations have been favoured. These are often linked to the new coalfield developments, such as the super pits in the Vale of Belvoir.

New conventional power stations cause environmental conflicts because of:

**66 Sources of conflict. 99**

a) their size which intrudes into the landscape, with the chimney rising high above the level surroundings;
b) the global environmental threats which have caused 'green' groups to protest about acid rain and pollution, although power stations will be cleaner in the future;
c) the loss of land, often in areas with valued ecosystems – Fawley is adjacent to marshland and close to the New Forest; and
d) disruption caused by construction traffic.

Nuclear power stations were the much vaunted cheap energy supply of the future. In the 1950s and 1960s the UK embarked on an ambitious programme of power stations to provide 20 per cent of the electricity supplies. While new stations are being built, such as Sizewell B at Leiston in Suffolk, using the later PWR (pressurised water reactor) technology), the older Magnox stations such as Berkeley in Gloucestershire (1962) are now being decommissioned and future developments have been curbed in the UK. Perhaps you could try to say why the opposition to nuclear power has grown but why governments are still (or have been) anxious to promote nuclear energy. What are the arguments for and against further power stations?

In November 1988, the AEB Paper 3 had an extract from Sheet 156 of the Ordnance Survey Second Series which showed the site of Sizewell. The part of the question which applied to the power station was as follows:

> The location of the Sizewell A nuclear power station is marked at 475632. The new Sizewell B nuclear power station, approved in 1986, is to be built immediately to the north of the existing nuclear power station in grid square 4763. What is the map evidence to suggest that the chosen site is a good one for the construction of Sizewell B?

**66 A possible exercise. 99**

You could get a copy of Sheet 156, answer this question and possibly go on to ask what form any local opposition to the station might have taken. Fig. 5.2 is a sketch of the layout. What case would the Friends of the Earth, the Town and Country Planning Association, the Campaign for Nuclear Disarmament, the National Union of Miners and the Central Electricity Generating Board make?

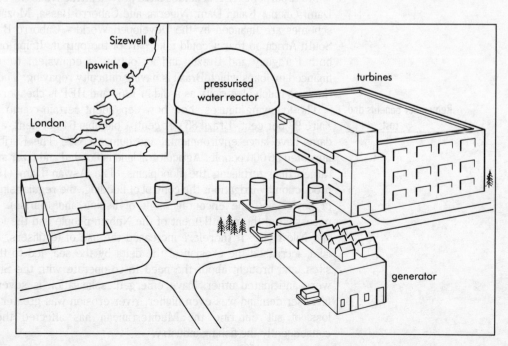

Fig. 5.2 Location of Sizewell and layout of Sizewell A

Hydro-electricity schemes, which use the energy of falling water, are also fraught with problems of environmental conflict. HEP has been most developed in the alpine states of Europe, e.g. Switzerland and Austria, and has become the major source of electricity for these countries. The factors affecting location are summed up in Fig. 5.3. However, the pressure to build further dams in Austria has met with opposition from those who feel that storage dams and power stations spoil the beauty of alpine valleys and affect ecosystems. More recently, environmentalists managed to prevent the building of a multi-purpose scheme on the River Danube at Hainburg near Vienna which would have provided power and regulated waterflow to aid shipping on the river. The reason for the opposition was that Hainburg is a unique wetland ecosystem, and conservation eventually overrode the need to provide more electricity. (Environmentalists even forced the Austrian government to abandon a completed nuclear power station at Zwentendorf before it was commissioned.) In both cases the environmentalists argued for more **energy conservation** as the alternative strategy to environmental degradation.

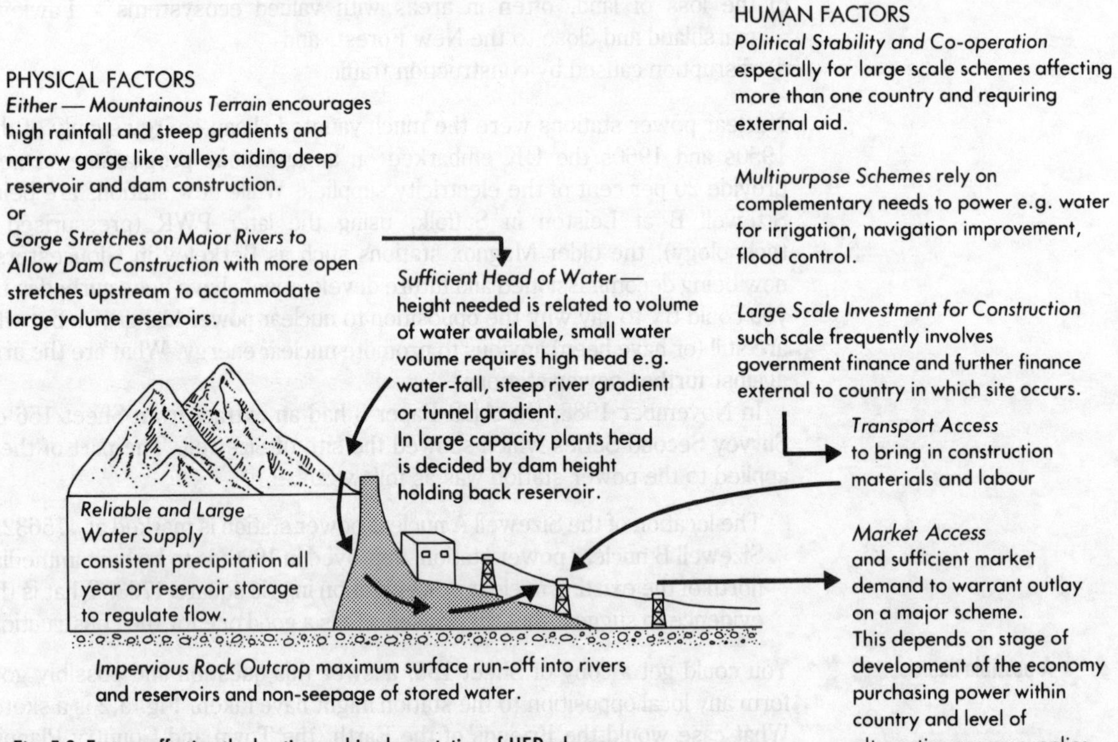

**PHYSICAL FACTORS**

*Either — Mountainous Terrain* encourages high rainfall and steep gradients and narrow gorge like valleys aiding deep reservoir and dam construction.
or
*Gorge Stretches on Major Rivers* to Allow Dam Construction with more open stretches upstream to accommodate large volume reservoirs.

*Sufficient Head of Water* — height needed is related to volume of water available, small water volume requires high head e.g. water-fall, steep river gradient or tunnel gradient.
In large capacity plants head is decided by dam height holding back reservoir.

*Reliable and Large Water Supply* consistent precipitation all year or reservoir storage to regulate flow.

*Impervious Rock Outcrop* maximum surface run-off into rivers and reservoirs and non-seepage of stored water.

**HUMAN FACTORS**

*Political Stability and Co-operation* especially for large scale schemes affecting more than one country and requiring external aid.

*Multipurpose Schemes* rely on complementary needs to power e.g. water for irrigation, navigation improvement, flood control.

*Large Scale Investment for Construction* such scale frequently involves government finance and further finance external to country in which site occurs.

*Transport Access* to bring in construction materials and labour

*Market Access* and sufficient market demand to warrant outlay on a major scheme. This depends on stage of development of the economy purchasing power within country and level of alternative energy supplies available.

Fig. 5.3 Factors affecting the location and implementation of HEP schemes

Multi-purpose HEP schemes have proved attractive in the Developing World (e.g. Volta Dam, Ghana; Kainji Dam, Nigeria; and Caborro Bassa, Mozambique; all in Africa). Most schemes are financed by the Developed World – Caborro Bassa was partly financed by South Africa so that it could take part of the output. Itaipu on the Parana River supplies both Paraguay and Brazil, and its output is equivalent to 13 nuclear stations. It was financed by loans which Brazil is having difficulty repaying. The Developing World uses six times as much electricity as it did in 1967 and HEP is cheaper than oil or coal power.

The following figures show how dependent certain countries are on hydro-electricity: Zaire 97 per cent, Brazil 85 per cent, Columbia 64 per cent, and Kenya 62 per cent. New dams have large environmental and human costs. The Kariba Dam, built in the 1960s, uprooted 56 000 people. Agricultural land is flooded and river silt is trapped behind the dam rather than fertilising the floodplains. The Aswan Dam (1970) was built to aid food production by irrigation, the control of flooding, the regularising of riverflow, the provision of power and the encouragement of fishing and tourism. However, its construction necessitated the resettlement of the Nubian people and the loss of silt in the Lower Nile and Nile Delta. It therefore increased the use of fertilisers, caused the silting up of the lake, increased the erosion of the delta by the sea, led to the moving of archaeological sites, and brought about the need to co-operate with the Sudan. Since these problems were anticipated others have emerged: salinisation is severe in some irrigated lands, fertiliser demand was even higher, river erosion was greater with the new flow, and the loss of silt entering the Mediterranean has affected the stocks of pilchard and, consequently the fishing industry.

Remember benefits and costs.

## ENERGY IN LESS DEVELOPED COUNTRIES

Energy is needed (1) for economic development and (2) for survival, providing fuel for cooking and heating, especially among the poorest people. The Developing World's sources of energy are:

1   The normal commercial fuels;
2   Biomass fuels which are not traded but are the key to survival;
3   Renewable sources such as wind, wave, solar, ethanol, biogas and tidal power.

**Energy for development** is generally imported and therefore exports are necessary to pay for these imports. Growth of industry necessitates an energy transition which is difficult to finance if oil prices rise faster than the country's exports.

**Energy for survival** is fuel for basic needs such as cooking. Timber for fuel is scarce so people have to spend more time going further afield to gather wood. Lesotho and Nepal are among the worst-affected countries.

Fuel conservation can take many forms:

> **Forms of fuel conservation.**

a)   Careful management to use less fuel on fire;
b)   Travelling further to gather fuel – affects other activities, usually of women;
c)   Adapting cooking to be communal, and therefore less frequent;
d)   Switching to crop and animal residues, but this generally occurs as a response to no supplies;
e)   Moving to kerosene and gas which are costly;
f )  Initiating the growing of timber for sale;
g)   Developing biogas from animal dung, using underground pits in which the dung ferments (China has 7 million such plants which provide both gas and organic fertiliser).

## THE POLITICS OF ENERGY

OPEC or the Organisation of Petroleum Exporting Countries controlled 65 per cent of the non-communist world production, and quadrupled prices in 1973. This precipitated an economic recession. Energy-short countries had to maintain good relations with the producers in order to maintain supplies. This further encouraged countries outside this cartel to explore for oil, e.g. the North Sea produced 6 per cent of world oil by 1984 and Mexico produced 7 per cent by 1984. During the Iran–Iraq war, countries took measures to ensure supplies and even assisted the combatants to maintain supplies.

The enthusiasm of the Thatcher government for nuclear power has been interpreted by some as a deliberate attempt to break the power of a trade union, the National Union of Miners, by ensuring that the UK can draw on a variety of energy resources. Governments will alter their energy policies if their majority is slim and if the issue might lose them votes. The Austrian examples in this chapter were the result of a strong 'green' lobby on a government which was losing popularity. Similarly the 'green' bandwagon has accelerated the programme of installing 'scrubbers' to remove sulphur from British power-station emissions.

> **Understand a variety of viewpoints.**

You should use newspapers to build up your examples of the politics of energy. Often you will have to decide which opinion fits in with your way of thinking. But do make sure that you understand the *basis* of the conflicting opinions which you might read. For instance, is a diversified energy policy for the UK a more important explanation of the policies pursued than that of crushing a trade union? You have to decide. What is crucially important is that you can use evidence and illustrations to support your arguments.

# EXAMINATION QUESTIONS

### Question 1

Study Fig. 5.4 which shows the limits to growth model of population development on a global scale. [The model assumes no major change in the relationships over time. All values until 1970 are based on reality, after 1970 the values are projected.]

a)   i)   Describe the form of:
       1   The population curve
       2   The resources curve                                                                 *(4)*
     ii)   Attempt to explain the relationship between the two curves.                          *(6)*

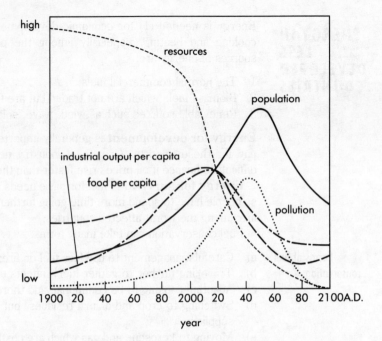

Fig. 5.4 A model of the limits to
growth on a global scale
*Source*: D. H., Meadows, D. C.
Meadows, J. Renders, and W. Behrews
*Report for the Club of Rome – Project
on the Predicament of Mankind*
(Universe Books, NY, 1972)

b) Give reasons for the predicted decline in industrial output per capita after AD 2020.　(6)

c) Suggest how the predicted global decline in food per capita might be arrested by technological innovation.　(5)

d) Explain the form of the pollution curve.　(4)

(London)

## Question 2

Study the figure below:

Fig. 5.5 World coal reserves

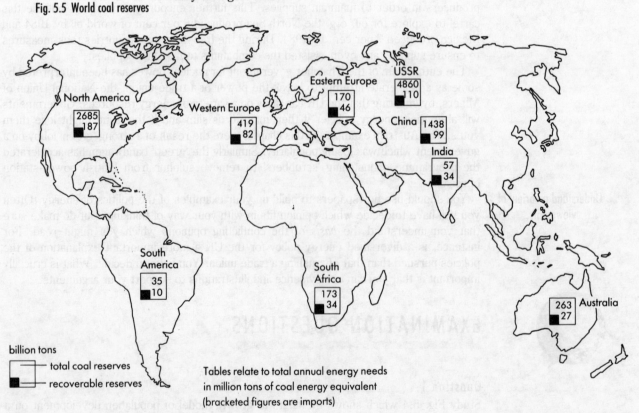

billion tons

▫ — total coal reserves
■ — recoverable reserves

Tables relate to total annual energy needs
in million tons of coal energy equivalent
(bracketed figures are imports)

| North America | | Western Europe | | China | | Japan | | Australia | |
|---|---|---|---|---|---|---|---|---|---|
| Oil | 1525 (705) | Oil | 1085 (440) | Oil | 110 | Oil | 430 (424) | Oil | 49 (16) |
| Nuclear | 150 | Nuclear | 90 | Nuclear | — | Nuclear | 30 | Coal | 46 |
| Coal | 545 (2) | Coal | 402 (100) | Coal | 380 | Coal | 90 (17) | Other | 22 |
| Other | 945 | Other | 302 | Other | 30 | Other | 60 | | |
| | | | | | | | | | |
| *Totals* | | *Totals* | | *Totals* | | *Totals* | | *Totals* | |
| 1980 | 3165 | 1980 | 1879 | 1980 | 520 | 1980 | 610 | 1980 | 117 |
| 2000 | 4120 | 2000 | 2636 | 2000 | 2000 | 2000 | 1030 | 2000 | 250 |

a) Comment critically upon the cartographic techniques used to construct this map. (5)

b) Describe and assess the implications of the distribution of coal reserves indicated on the map. (10)

c) *Either:*
   i) Assuming that by the year AD 2000 coal will be supplying a significantly higher proportion of energy needs, assess the likely environmental effects of increasing reliance on coal as a source of energy. (10)

   *Or*
   ii) What energy sources might be considered under the heading 'other' and what is the likely environmental impact of their exploitation? (10)

(London 16–19, Sample AS Paper)

## Question 3

What have been the principal consequences for the production, distribution and consumption of energy in the UK of the trends indicated in Fig. 5.6? (25)

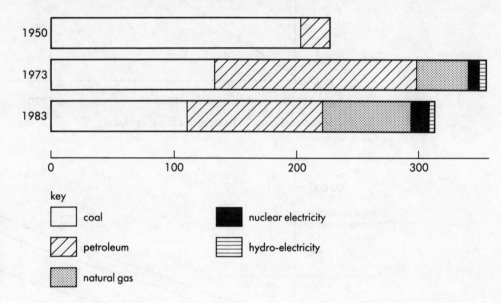

Fig. 5.6 Sources of UK inland energy consumption, 1950–83 (in million tonnes of coal equivalent)
*Source: Annual Abstract of Energy Statistics*

## Question 4

Analyse the geography of world petroleum with regard to (a) production and (b) consumption.

(Oxford, Sample AS Human Geography)

## Question 5

a) Compare the types, location and relative importance of the energy resources of Nigeria, Egypt and Tanzania. (10)

b) Discuss the ways in which the use of the energy resources of TWO of these countries has influenced their economic development. (15)

(JMB 'C' Paper II, Alternative)

## Question 6

Study Figs 5.7, 5.8, 5.9 and 5.10, which show information about energy production and consumption for selected countries in Central and South America.

a) Compare and contrast the countries in terms of the changes in their total energy production as shown in Fig. 5.7. (7)

b) Suggest reasons for the differences between the trends of energy production shown in Fig. 5.7 and the trends of energy consumption shown in Fig. 5.8, both within and between individual countries from 1950 to 1980. (8)

c) Discuss the changes in the proportions of different sources of energy produced and consumed in Brazil from 1950 to 1980 as shown in Fig. 5.9 and 5.10. (10)

(JMB 'C', Alternative, 1986)

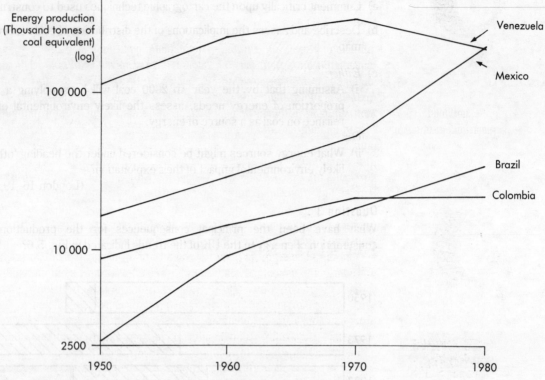

Fig. 5.7 Total production of energy (in thousand tons of coal equivalent)

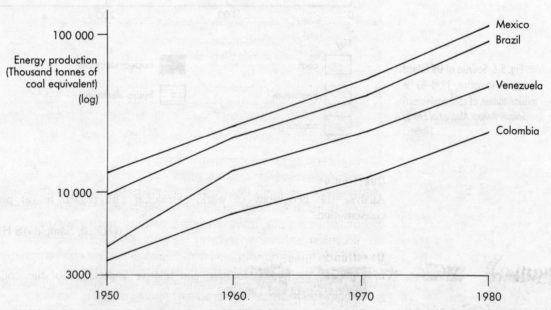

Fig. 5.8 Total consumption of energy (in thousand tons of coal equivalent)

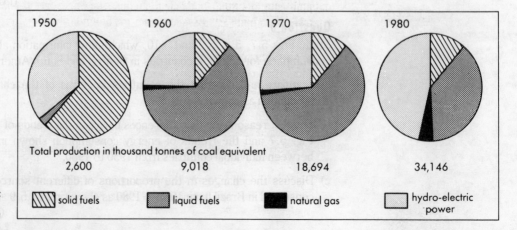

Fig. 5.9 Brazil: production of
energy by type

Total production in thousand tonnes of coal equivalent

2,600          9,018          18,694          34,146

solid fuels          liquid fuels          natural gas          hydro-electric power

Fig. 5.10 Brazil: consumption of energy by type

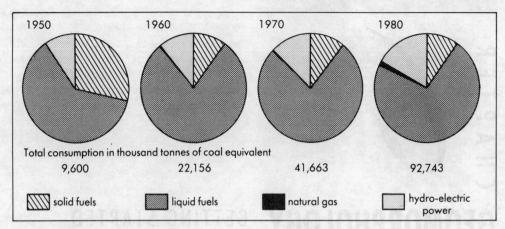

1950          1960          1970          1980

Total consumption in thousand tonnes of coal equivalent
9,600         22,156        41,663        92,743

solid fuels       liquid fuels       natural gas       hydro-electric power

# OUTLINE ANSWERS

## Answer to Question 1

a) i) 1 Population rising to a peak in 2050 and falling more rapidly until 2100.

   2 The resources curve declining gradually until 2000 then an exponential decline over the next century.

   ii) There is an inverse relationship: as population rises so resources fall. Once the level of resources is unable to support the population, then the population declines in a Malthusian fashion.

b) (At least 2 reasons needed)

   1 Decline in resources for industry.

   2 Economies depend increasingly on other activities.

   3 More people, therefore output per head will decline.

   4 Total output is static.

c) By new strains of crops producing more.

   By increased double-cropping and irrigation.

   By use of improved techniques such as hydro policy.

d) The rise coincides with and reflects the growth of population, but as population and resources decline there will be less consumption to cause pollution. Possibly some better control over pollution as well.

## Answer to Question 2

a) By failing to use the Peters projection, the land areas of the respective countries/continents are distorted. Perhaps too many variables are displayed on one map: coal reserves, annual energy needs for various fuels, imports of various fuels.

b) Use the figures and tables provided to show that coal reserves are on the whole, evenly distributed throughout the world – more so than oil, say! Look at the figures for recoverable coal reserves in the *boxes* and those for annual energy needs for coal in the *tables*. Remember that the box figures are *billion* tons but the table figures are *million* tons. Use your data to show the considerable number of years of recoverable coal reserves available for the various countries. Still more so if new technology makes a higher proportion of the known reserves recoverable in the future, etc. You could also look at the implication of the distribution of coal reserves for the balance of payments – e.g. note figures in brackets in tables (coal imported) as a proportion of annual needs for coal for various countries/regions.

c) Here we look at answer (i).

   i) You might start by using the data in the map to estimate the *current* proportion of energy needs supplied by coal. We are now told that this proportion is to rise significantly. You could look at various effects, such as the increased sulphur dioxide emissions from power stations using coal, housing and heating systems using coal, etc. This is likely to increase the amount of acid rain, resulting in deforestation, reduced fish stocks, depletion of ozone layer, etc. etc. Need for expansion of existing mines/new mines has impact on landscape, local communities, occupational profile, etc.

   However, remember increased coal use may imply reduced (relative) use of *other* energy sources – perhaps nuclear! – so offsetting environmental gains might be achieved.

# GETTING STARTED

Physical geography is a very broad area of the syllabus which requires much knowledge and understanding. You should be aware of how physical processes which you have studied (1) affect landforms, (2) affect the economic and social activities of societies and, conversely, (3) how people may modify those physical processes, and (4) with what effect.

In Geomorphology we are concerned with landforms, the processes which have shaped them and the consequences for human activity of both the processes and the forms.

# ESSENTIAL PRINCIPLES

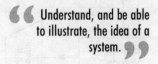

**BASIC THEMES IN GEOMORPHOLOGY**

### System

'System' is a key term in many areas of Geography – it is an abstraction of the real world with a structure which functions in a particular way as a result of the flow and transfer of energy and material. **Environmental systems** are **open systems** with continual throughputs of energy and matter.

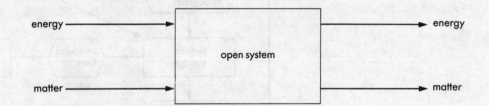

> Understand, and be able to illustrate, the idea of a system.

Fig. 6.1

Fig. 6.2 illustrates one system: the catchment basin.

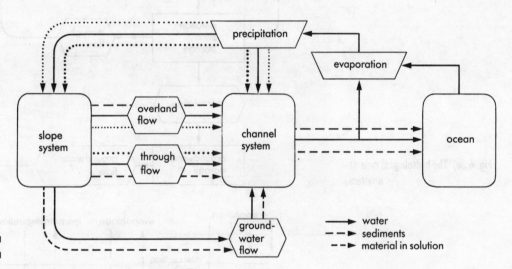

Fig. 6.2 The catchment basin system

### Lithosphere

The lithosphere is composed of the rocks of the earth's crust which make up the seven major and twelve minor plates and are subjected to the forces of weathering and erosion. It provides nutrients for the **biosphere** (chapter 8).

### Hydrosphere

The hydrosphere is a further component of the physical environment which relates to the surface water and is linked to the atmosphere (chapter 7) by the hydrological cycle. The **biosphere** transfers matter from the hydrosphere to the atmosphere.

### Plate tectonics

This theory was developed in the 1960s to explain the pattern of the earth's components. The plates are moved by convection currents in the asthenosphere, are bounded by **plate margins** which can be **constructive, destructive** or **conservative**. It is the theory which helps explain the patterns of earthquakes, volcanoes and other structures such as **oceanic ridges and trenches**, and island arcs.

### Hydrological cycle

The hydrological cycle is the movement of water through the atmosphere, biosphere and lithosphere (Fig. 6.3).

### Weathering

Weathering is the process by which rocks at or close to the surface are broken down and/or decay. It is the early part of **denudation**. It is subdivided into **chemical, mechanical** and **organic**, and leads to the production of **regolith**, the decomposed rock material and, as a result of **pedogenic processes, soil** (chapter 9).

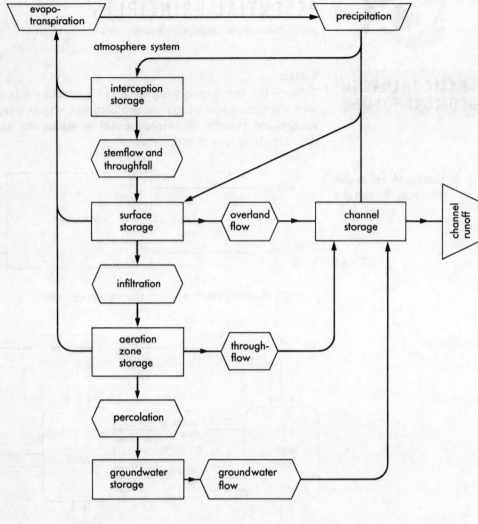

**Fig. 6.3a)  The hydrological cycle as a system**

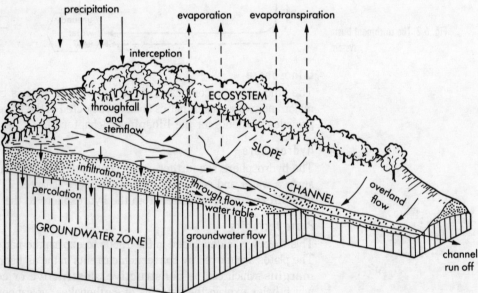

**Fig. 6.3b)  The hydrological cycle for groundwater flow**

### Erosion

The second stage of denudation is erosion – the physical processes shaping and moulding landforms as a result of the work of running water, sliding ice, breaking waves and wind-borne grit and dust. Chemical erosion can occur in the production of coastal scenery and **Karst** landscapes.

### Transport

The movement of material through a system from the place of erosion to the site of deposition is referred to as transport. In water, **saltation, solution, suspension** and traction move material. **Waves, tides** and **currents** operate in marine environments. **Glaciers** and **wind** also transport material. **Gravity** is a further agent on slopes. The particles moved or **load** vary in size and are **sorted** in transport. Further erosion of the particles, **attrition**, occurs during transport and the load erodes the channel through which it is moving by attrition.

### Deposition

The final stage of denudation is deposition, resulting in the laying down of sediments as the energy to transport the material declines. The results are landforms of depositions, for example, **drumlins, dunes, alluvial fans, berms**. Some depositional landforms are more permanent than others.

### Drainage basin or catchment basin

This is the basic functional unit through which the denudation system operates. It is bounded by **watersheds** and, more difficult to define, within the lithosphere. It can be comprised of sub-catchments for each tributary. Within it the **slope sub-system** and the **channel sub-system** remove weathered material. Streams within the system can be analysed in terms of **network characteristics** such as **stream order**, numbers per order and length per order. The **drainage density** of streams can vary with the seasons.

### Slope processes

Slopes are the fundamental unit of all landforms. Relationships exist between the **form** of the slope and the processes operating on it. The form is measured in terms of **angle, profile** and **depth of regolith**, which are the product of past processes and, in turn, influence present processes. Water is essential for the movement of material, together with the effects of gravity in **mass movement** through **slide, flow** and **heave**, to produce **soil creep, frost creep, rockslides, landslides, earthflows** and **mudflows**. Gravity alone affects only the most vertical slopes to give rockfalls. **Rainsplash, sheetwash** resulting in **rills** and movement in solution alter slopes. Models of **slope retreat** have been developed.

### Natural hazards

Many geomorphological events are seen as hazards because of their threat to human life. This is especially the case with **volcanic activity, earthquakes, floods** and catastrophic events on slopes such as mudflows. On the other hand, there are benefits which can stem from hazard events such as volcanic soils and new alluvium on floodplains.

### People and landforms

Landforms can be the product of human interference in delicately balanced processes of slope formation. The increase of erosion due to cropping, leaving soils exposed, is one case. The effects of building **groynes** on beach formation and the subsequent loss of sand is another unwitting interruption of a process, brought about by human interference. **Sea walls** to protect cliffs, **artificial levées** and **dune stabilisation** are other examples of human modification of processes of erosion, transport and deposition.

Remember that the People and Environment theme of many syllabuses does lay stress on the effects of natural processes on economic and social activities and, conversely, the effects of people on the natural processes.

## THE STRUCTURE OF THE EARTH

66 *Many questions make use of a knowledge of plate tectonics.* 99

### Plate tectonics

The earth's crust or lithosphere averages 33 km in thickness in continental areas and 11 km in oceanic areas. The lithosphere rests on the plastic zone of the **mantle** or **asthenosphere** which extends to about 250 km in depth. **Continental crust** is mainly granitic whereas **oceanic crust** is basaltic and younger. These crusts form the major plates. **Mid-oceanic ridges** are found in all the oceans associated with high convective heat flows from the **mantle**. New crust is being created along these ridges as the oceanic plates are pushed apart by convection currents. Oceanic margins are characterised by **oceanic trenches** where the crust slides beneath the continental crust along the **Benioff zone**. It is the shearing of the descending plates which gives rise to earthquakes and **volcanic** and **geothermal activity** on the surface.

## Plate margins

These are **constructive** or **extrusion zones** where material is added, and **destructive** or **subduction zones** where material is consumed, e.g. the Andes. They are **transform** where plates slide laterally, e.g. San Andreas fault.

Destructive margins (Fig. 6.4) can also occur where two ocean plates collide, giving rise to **island arcs**, and where two continental plates collide, as along the Himalayas.

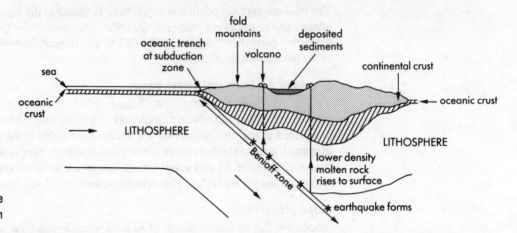

Fig. 6.4 The destructive plate margin

## Earthquakes

Earthquakes are produced by sudden movements of the earth's crust which create a series of shock waves or **seismic energy**. The **focus** or **epicentre** is normally associated with plate margins. The strength is measured on the **Richter Scale**, a logarithmic scale, so that 6.0 is ten times stronger than 5.0. If they occur at sea, then sea-bed movement can cause **tsunamis**, huge waves which will flood low-lying areas. Osaka has built large flood gates to the harbour and sea walls to counteract flooding by tsunamis. Earthquakes are a major hazard (see the section on Environmental Hazards, below). Direct hazards include **liquifaction** so that solid ground appears to shake like a jelly, and building destruction. Indirect effects include fire (700 died in San Francisco in 1906), ruptured water and sewage pipes, breaking dams, damage to bridges, landslides, flooding and crop destruction.

Earthquake monitoring has increased: special regulations may govern building height, structure and densities. Experiments to lessen the shock by injecting fluid into the fault have been attempted.

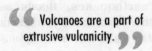
Volcanoes are a part of extrusive vulcanicity.

## Volcanoes

These are a part of **extrusive vulcanicity** and are mainly associated with plate margins or former margins. They are **extinct**, e.g. Puy de Dome, **dormant** (the distinction is hard to prove), e.g. Kilimanjaro, or **active**, e.g. Etna. Mount St Helens was dormant for a century before the 1980 eruption. There are two types of eruption: (1) the **Hawaiian** which is least violent and produces a **sliced volcano** from the basic lavas of the oceanic crust, and (2) the **Plinian**, such as Vesuvius in AD 79, creating a large cone with acid lava from continental crust. A **caldera** is a modified cone caused by a massive explosion, e.g. Crater Lake, Oregon. **Hot spots** in the mantle give rise to volcanic areas in the middle of plates, e.g. Hawaiian islands where the spot has moved to produce a series of volcanic islands.

Material from eruptions, **scoria** or ash dust, **lapilli** and **tephra** can travel great distances, even into the upper atmosphere.

Volcanoes and associated features are tourist attractions, e.g. Etna, the **geyser** 'Old Faithful', boiling mud, sulphur springs, and a 'walk-in' volcano in St Lucia. The soils derived from extrusive vulcanicity are often fertile and therefore attract people to hazardous areas, e.g. Sumatra. Volcanic activity can threaten livelihood, e.g. Vestmannaeyjar harbour in Iceland in 1973, or slowly destroy a town, e.g. Pozzuoli, as the area gradually rises from the sea.

Extrusive vulcanicity can also result in **lava flows**, e.g. the Deccan, and **fissure eruptions**.

## Intrusive vulcanicity

Intrusive vulcanicity occurs when **magma** forces itself into the country rock but does not reach the surface. Subsequent erosion often exposes the intrusions of **dykes**, vertical

**sills**, horizontal **intrusions, laccoliths** – small domes – and **batholiths** – large domes like Dartmoor. The Great Whin Sill at Bamburgh provided a defensive site, whereas the Dartmoor batholith with its upland moorland granite scenery is the focus of tourism, military training and hill farming.

We can now look at some of the key processes in geomorphology.

## WEATHERING

Weathering is the breakdown of rock at or near the surface and is the first stage of the denudation process.

### Chemical weathering

Chemical weathering is a series of reactions leading to the breakdown or decay of rocks.

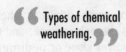
*Types of chemical weathering.*

- **Carbonation** – dominant in Karst (see below). Water + $CO_2$ forms weak carbonic acid ($H_2CO_3$) which can act on limestones, carrying the dissolved rock away as calcium bicarbonate.

- **Hydrolysis** – the dominant type of weathering, where water and rock minerals combine to form insoluble clay minerals, except carbonation (above) where there is a soluble product.

- **Oxidation** – the oxygen in water reacts with rocks containing iron to form brownish oxides and hydroxides.

- **Reduction** – the removal of oxygen due to continuous presence of water associated with **gleying** (chapter 9).

- **Solution** – relatively unimportant: the dissolution of rock by water, most noticeable in Karst.

### Mechanical weathering

Mechanical weathering is the break up of rocks caused by:

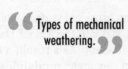
*Types of mechanical weathering.*

- **Exfoliation** – the splitting off of the outer layer of rock by the expansion of salt crystals in the surface layers and by heating of the outer shell.

- **Freeze–thaw** – water freezes in fissures, splitting rocks by increased pressure, then thaws and refreezes, so weakening rock.

- **Granular disintegration** – caused by either the freezing of pore water or by expansion and contraction due to **insolation**, resulting in break-up of rock.

- **Sheeting** – the splitting off of the upper layer of rock once exposed by denudation: normally found in igneous rocks.

### Organic weathering

Organic weathering is the breakdown of rocks by plants (flora) and animals (fauna). Tree roots and burrowing animals can break up rocks. **Decomposition** creates organic acids which chemically decompose the rocks, and affects stone in buildings, which can be exacerbated by acid rain. The weathered material is removed by the processes of erosion acting on the surface of the slope.

## SLOPES

Slopes are generally classified by their **form**. Most have four units: **waxing** (convex) slope, **free-face**, **constant slope** and **waning** (concave) slope (Fig. 6.5). Some add a fifth unit or **talusfoot** between the third and fourth elements in mountain areas, and there is a nine-unit model.

### Models of slope evolution

These models are *not* mutually exclusive.

- **Davisian Slope Profile Decline**: a balance between weathering and transport results in a declining angle of slope. At points on the lower profile transport has to move all the upper material transported down to that point, plus the weathered material at that point. Does seem to characterise older phases of landscape development.

- **Penckian Slope Replacement**: a **cliff slope** will be replaced by a **debris slope** which in turn is replaced by a **foot slope**. The sequence repeats itself to produce gentler slopes.

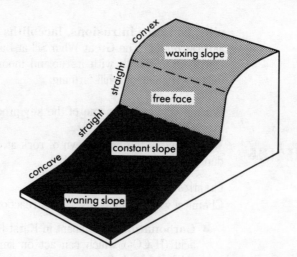

**Fig. 6.5 The slope profile**

■ **King's Parallel Slope Retreat:** characteristic of many semi-arid areas when the cliff face is always the same proportion of the slope as is the debris slope, and these retreat in parallel, leaving an increasingly large concave lower slope or **pediment**. It is a very common slope form of **mesa** and **butte** landscapes.

### Factors influencing slope development

1 Rock type;
2 Angle of dip;
3 Climate;
4 Human activity.

### Movement on slopes

Mass movement can be classified by process and material.

■ **Heave** – the alternating expansion and contraction of loose rock pieces which move downslope. **Creep** is controlled by slope angle and can affect all the weathered surface. Movement is greater near the surface. **Solifluction** is more common in periglacial areas. Result in terracettes and has effects on walls and trees.

■ **Slide and flow.** Sliding occurs over bedding planes or an erosional surface which is lubricated by water. In 1963, 600 million tonnes of rock slid into the Vaiont Dam in North Italy, forcing the water to rise 100 m above the dam crest. As a result, a flood sped at 30 m/sec down the valley, killing 2500 people. In clay areas **mudslides** and **mudflows** are more common. **Sheetflow** occurs mainly on unvegetated angle slopes after heavy rain and moves particles down slopes, sometimes concentrating the flow into **rills**.

■ **Rotational slips** – classically associated with permeable rocks overlying impermeable rocks. The angle of dip of the strata and the presence of water to act as a lubricant aid movement along a **slide plane**. Frequently associated with marine cliffs where removal of material at the base is more rapid.

■ **Slumps** occur where the coherence is lost due to saturation, and the rock body moves along a **slide plane** reaching a flow-like state at the **toe**. Slumping can frequently be seen in recent motorway cuttings, especially through clays, where the angle of rest of the cutting is too steep for saturated conditions.

■ **Factors influencing mass movement**
   1 Internal rock pressure;
   2 Opening of joints;
   3 Gravity slope angle – steep enough;
   4 Material which can be put in motion;
   5 Water, heavy rainfall;
   6 Human action – deforestation, ploughing before planting, cuttings, quarries;
   7 Earthquakes can trigger flows especially if there has been heavy rainfall.

## RIVER CHANNELS – THE FLUVIAL SYSTEM

The channel contains water from **overland flow, groundwater flow, throughflow, precipitation** and from other tributaries, which is carried to the oceans or an inland sea or lake. It is measured by the hydrograph – one for a flood is shown in Fig. 6.6.

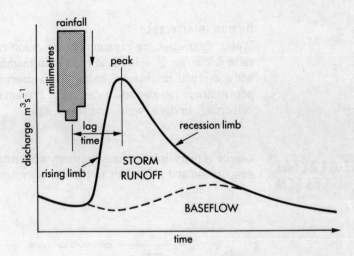

Fig. 6.6 The storm hydrograph

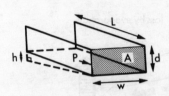

Fig. 6.7 The channel variables

## Channel variables

**Width (w)** – measured at water surface;

**Depth (d)** – varies, but normally expressed as mean depth across section (J);

**Cross-section area (A)** – Jw or mean depth × width;

**Wetted perimeter (P)** – length of channel margin in contact with flowing water
P = 2d + w;

**Channel gradient (S)** – change in elevation (h) per unit of length (L);

**Velocity of flow (V)** – distance travelled per unit of time;

**Discharge (Q)** – volume of water passing through a cross-section per unit of time.

The width, depth and velocity all change with discharge and these relationships may be illustrated for any stream.

## Transport of load

Transport of load is as **bedload, suspended load** and **solution load**, and the amount of load depends on the energy of the stream, i.e. velocity: floods are therefore more effective at moving load. The nature of the material adjacent to the stream will also affect the load.

## Deposition

As discharge declines, particles are redeposited as gravel spreads and **braiding** or as **point bars** inside meanders or as alluvial sediment after a flood has subsided. Deposition also takes place in lakes.

## Channel form

**Long profiles** are concave. As drainage area and discharge increases downstream so cross-sectional area increases to deal more efficiently with increasing discharge and to dissipate energy by friction with the sides.

## Channel plan

Channel plan is generally sinuous, i.e. meandering. The characteristics of a meander are shown in Fig. 6.8. Braiding into **distributaries** may occur, especially in deltas.

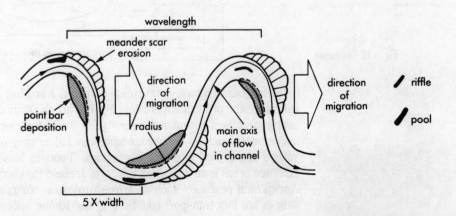

Fig. 6.8 Meander terminology

### Human interference

Water abstraction for consumption/irrigation can affect discharge; similarly, effluent will raise discharge. People can also affect throughflow and groundwater flow, and urbanisation will accentuate discharges, owing to increased overland flow. Vegetation and landuse (e.g. afforestation) can also affect discharge. Stream geometry and discharge may be affected by culverting, bridges, weirs and reservoirs.

## THE GLACIAL SYSTEM

Glacial activity is a balance between **accumulation** of snow and **ablation** or loss by evaporation and meltwater to the fluvial system (Fig. 6.9).

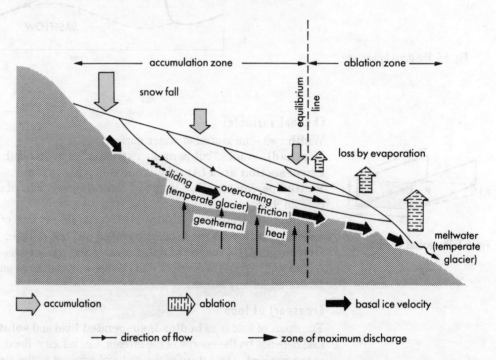

Fig. 6.9 The glacial system

Glaciers flow by **basal sliding**, as seeping meltwater accumulates at the base, and **internal deformation**, when the weight of the upper layers of ice causes the lower layers to behave like a fluid. The flow varies with nature of the bedrock, amount of ice, and loss of ice.

### Glacial erosion

**66** Aspects of the glacial system. **99**

Glacial erosion takes place by **abrasion** and **plucking**. It is influenced by rock structure and the relief of the area. The most distinctive landform is a cirque (Fig. 6.10).

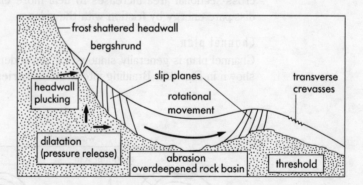

Fig. 6.10 The cirque

In the glaciated valleys **truncated spurs, hanging valleys, rock steps** and **ribbon lakes** are the products of erosion. **Fjords** are the distinctive coastal landform. The eroded landscape offers potential for dam sites for hydro-electricity, often to supply power to major users such as aluminium smelting. Lakes may also provide water supply and the valleys are typical sites for reservoirs. Tourism based on the spectacular scenery of glaciated areas is an important use, e.g. Icefield Parkway in the Alberta Rocky Mountains. Agricultural practices such as **transhumance** are generally in decline. Many glaciated valleys are key transport routes, e.g. the Rhône valley in Switzerland.

## Glacial deposition

Glacial deposition is generally called **drift**; it can be **glacial**, **fluvioglacial**, **lacustrine** or **marine**. Deposition occurs beneath the glacier, by dumping at the margins, and by **melt-out**.

The most widespread feature is **till sheet**, e.g. over East Anglia, whose origins can be determined by **till fabric analysis**. **Moraines** are of varying types depending on their location, i.e. **terminal**, **push**, **recessional** and **side**. **Drumlins** (Fig. 6.11) i.e. parallel to the direction of ice movement are 100 per cent till. **Crag** and **tails** develop where rock outcrops obstruct glacial flow. **Erratics** provide clues to the origin of post-glacial activity in an area.

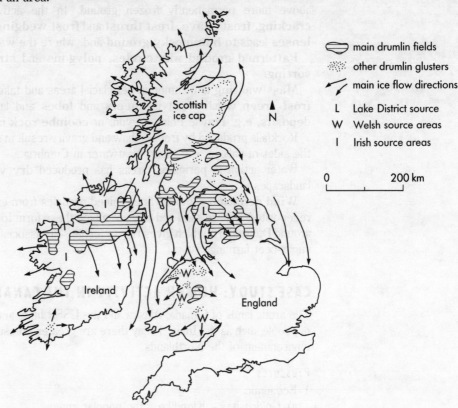

main drumlin fields
other drumlin glusters
main ice flow directions
L   Lake District source
W   Welsh source areas
I   Irish source areas

0          200 km

**Fig. 6.11 The origins and distribution of drumlins**

East Anglia and much of northern West Germany (GFR) and East Germany (GDR) have landscapes produced by glacial deposition.

## Glacial meltwater

Glacial meltwater comes from the surface, the valley sides and beneath the glacier. A model of this activity in system terms is shown in Fig. 6.12. Meltwater can result in the formation of lakes which have since disappeared, e.g. Lake Harrison in the Stratford to Leicester area, leaving only evidence in the form of **strand lines** and **varved clays**. Meltwater often forms **channels** such as those on the North York Moors, e.g. Newtondale, or the large Urstromtäler on the North German plain which are used by the inland waterways to link Berlin to the Ruhr.

**Fig. 6.12 The meltwater system**

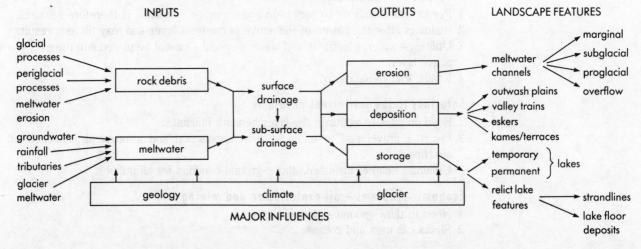

INPUTS            OUTPUTS            LANDSCAPE FEATURES

glacial processes
periglacial processes          rock debris
meltwater erosion                          surface drainage          erosion          meltwater channels          marginal / subglacial / proglacial / overflow

groundwater
rainfall          meltwater
tributaries
glacier meltwater                          sub-surface drainage          deposition          outwash plains / valley trains / eskers / kames/terraces

                                   storage          temporary / permanent } lakes
geology          climate          glacier          relict lake features          strandlines / lake floor deposits

MAJOR INFLUENCES

**Meltwater deposition** is most frequently in the form of **meltwater plains**, which may contain **kettle holes**. On these plains **eskers**, **kames** and **kame** terraces develop. The North German plain has mainly fluvio-glacial deposits which lie upon the glacial tills and have given rise to heathlands, e.g. Luneburg.

### Periglaciation

Aspects of periglaciation.

Periglaciation occurs in areas with a cold but non-glacial climate and therefore the effects may be seen today in, for example, the Canadian Northlands, or in relic form in areas which once fringed glaciers, e.g. the North Downs.

**Permafrost** is a zone of frozen ground and is comprised of a seasonally active layer above more permanently frozen ground. In the active zone **freeze–thaw**, **frost cracking**, **frost heave**, **frost thrust** and **frost wedging** take place. Ice in **wedges** and **lenses** leads to **hummocky ground** and, where the water volume is larger, **pingos**.

**Patterned ground** with **circles**, **polygons** and **stripes** develops owing to **frost sorting**.

**Mass wasting** is common in periglacial areas and takes the form of **solifluction** and **frost creep** which result in **sheets** and **lobes** and larger deposits known as **head deposits**, e.g. Prawle Point, Devon, or **coombe rock** in the South Downs.

Rockfalls produced by freeze–thaw and gravity result in **scree** which often characterises the sides of valleys, e.g. near Wastwater in Cumbria.

Water action in periglacial areas has produced **dry valleys** and **coombes** in chalk landscapes.

**Wind deflation** removes fine-grained particles from exposed surfaces which are then redeposited. Where sand and clay are mixed these form **loess**, e.g. the lower Rhine valley around Düsseldorf, which gives rise to fertile soils associated with major areas of grain and sugar-beet farming.

## CASE STUDY: HUMAN ACTIVITY IN THE CANADIAN NORTH

The arctic lands of Canada, Alaska and the USSR have always been home to small groups of people such as the Inuit. Today there are greater pressures being exerted on the fragile environment of the Northlands.

### Pressures

1 Economic
   a) Goldrushes – Klondike – the popular image.
   b) Mineral exploitation – iron-kimena or oil on the Alaskan north slope.
   c) Fishing (or whaling) resources – hunting the oldest.
2 Strategic
   a) The USA and USSR are on either side of the Pole and the Northlands represent a vital line of defence for both the Soviet Bloc and NATO.
   b) The need to control areas in case of mineral discoveries.

### Consequences

1 Settlements for mining, fishing, military ports, outposts.
2 Need for transport – airstrips, roads, pipelines, railways.

### Problems for settlements

1 Permafrost is difficult to penetrate and excavate – building is therefore seasonal.
2 Buildings alter the nature of the active permafrost layer and may tilt as a result.
3 Utilities – water, electricity and waste disposal – cannot be placed into the ever-moving active layer.
4 Roads, etc. move easily.

### Solutions to the settlement problems

1 Build on piles to ventilate the layer beneath buildings.
2 Insert a gravel pad to minimise heat conduction and improve drainage. Used for airstrips.
3 Utilidors: heated, insulated, above-ground conduits for all utilities.

### Economic problems – oil exploitation and mining

1 Need to thaw ground – slow process.
2 Shafts can melt and collapse.

3 Drilling mud can freeze.
4 Oil must be kept warm to flow up well through permafrost.
5 Pipeline must also be kept above ground and warm to permit flow.
6 Raised pipes may affect animal migrations.

## Solutions

1 Add brine/chemicals to drilling mud.
2 Insulate well/pipeline.
3 Pipeline in an insulated duct.
4 Pipeline zigzags across country so that expansion/contraction of pipes is easier.

This case illustrates very briefly how particular environmental conditions can affect the way people live.

## ARID PROCESSES

Arid areas are where there is a **permanent water deficit**: annual potential evapo-transpiration exceeds annual precipitation. These can be hot or cold areas. The cause of aridity is climatic (see chapter 7).

> Aspects of arid processes.

### Weathering

Weathering takes the forms of **insolation weathering** caused by diurnal temperature ranges and **freeze–thaw** in cold deserts, **chemical weathering** and **salt weathering**, especially in the pores of sandstones. Weathering produces rockwaste or **lithosols**, **zeugen** or **mushroom rocks**, and **gnamma holes** – pits hollowed out by weathering.

### Crusts

Crusts such as **desert varnish** develop from **capillary action**. They are formed of **calcretes, gypcretes** (Namib desert) and **silcretes** (Lake Eyre, Australia).

### Stone pavements

Stone pavements remain after finer materials are removed by the wind and **flash floods**. **Hamada** is a coarse stony landscape and **reg** a gravel-like desert.

### The role of water in deserts

Rainfall in intense, highly localised storms soon spreads as **sheet flood** and **stream flood**, encouraged by lack of vegetation, the presence of hard crusts and lack of infiltration. The streams are **ephemeral** and utilise **wadis**, valleys and **arroyos** (the gullies). There are flash floods which erode and transport material to deposit it as **alluvial fans** which may coalesce to form **bajada**. There are permanent streams, e.g. the Nile which is **exogenous**, or those which flow into a lake basin such as Lake Eyre, i.e. **endoreic**. Water is also present as dew which accelerates weathering.

### Wind (Aeolian) erosion and deposition

Material is removed by **deflation** which creates huge depressions. These may reach the water table, to produce an **oasis**. Wind-blown sand will sculpt rocks, often in alignment with the dominant winds to form **yardangs** or polish boulders known as **ventrifacts**. The sand particles are transported in a series of jumps called **saltation**, although some move by **surface creep**. The sand accumulates in large areas: **sand seas** or **ergs**. The sea is often formed of **seifs** or longitudinal dunes in long lines parallel to the dominant wind, e.g. in the Australian desert. **Barchans** are crescent-shaped isolated dunes, again aligned in a direction. These move forward and may become seif dunes.

Satellite photography has revealed larger scale depositional features. The **draa** are large ridges of sand which contain other dune forms. **Rhourd** is a star-shaped mound 1 km across × 150 m high, possibly formed by winds from differing directions.

Desert conditions affect human activity. In the previous sections on rivers and glaciation we outlined some effects. In this case, try to outline for yourself the effects of arid conditions on agriculture, industry, transport, housing.

## COASTAL PROCESSES

Coasts are the product of marine erosion and deposition by waves, tides and currents.

### Waves

Waves are formed by the movement of wind over water, and as waves reach shallow water

> **Waves are formed elsewhere and travel great distances.**

their size and shape changes until they break on the shore. Breakers **surge, spill, collapse** and **plunge** onto the shore to form **swash** up the beach and **backwash** as they return. Stronger swash is **constructive**, whereas stronger backwash is **destructive**. Swash and backwash interact in a complex fashion as each wave approaches, and complex **rip currents** can be set up, sometimes carrying unfortunate swimmers well offshore (many east Australian beaches can have severe rip currents running up to 1 km offshore). Waves rarely approach parallel to the beach – therefore **longshore drift** may occur and need to be controlled by **groynes**. Waves also **refract** to a pattern mirroring that of the coastline, so dissipating the wave energy in a bay but intensifying it on a headland.

### Tides
**Tides** vary in amplitude in different areas and at different times. **Tidal currents** are affected by the **range** and alter the transport of eroded material.

### Weathering
Weathering on cliffs will occur as on slopes, and slopes will move by **slumping, rotational slip** and even earth and mudflows.

### Erosion
Erosion occurs on *all* coasts, even those experiencing deposition. It is produced by wave attack undercutting the land and working upon the bedding planes and joints by **abrasion** and **hydraulic action**. Geology and especially **lithology** is most important. Hardness and exposure will determine the rate of retreat. As cliffs retreat, **blow holes** (geos), **gullies, arches, caves** and **stacks** may be formed due to **differential erosion** and **wave-cut platforms** are formed at the high or low or intertidal levels. The platform helps dissipate wave energy and slow down erosion (Fig. 6.13).

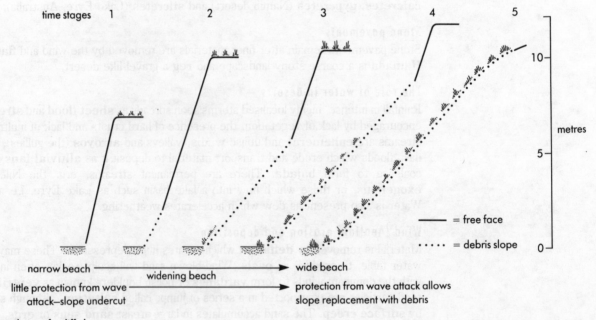

Fig. 6.13 The evolution of a cliff slope

Preventing cliff erosion is a very important task in areas of rapid cliff retreat, e.g. Happisburgh, Norfolk, and Fairlight near Hastings, because property has been and will continue to be lost as cliffs collapse.

Sea walls are designed to prevent such erosion but are expensive to build and maintain. They can also prevent deposition in other areas because the supply of material has been cut off.

### Deposition
**Beaches** are a store of sediments whose layout and composition vary with waves, weather, season, the type of sediment and human interference. They can be seen in systems terms (Fig. 6.14). The landforms of a beach are shown in Fig. 6.15.

**Spits** are very common depositional features in areas of small tidal range. Spits need a supply of material, either from longshore drift or movement of material from offshore. They build up as **lateral ridges** and have a **recurved** (hooked) end. Pairs of spits may form **cuspate forelands**, e.g. Dungeness, or link the mainland to an island – the tombolo

– or completely block a river estuary as a **bar** which may enclose a freshwater lagoon. Saltmarshes are often found in the lee of spits and develop a very distinct ecosystem, which is threatened by the use of spits for recreational purposes.

**Coastal sand dunes** depend on available dry sand, a surface to build on and a stabilising agent such as vegetation. **Embryo** and **fore dunes** are formed and these develop their own distinct ecosystems both on the dunes and in the **slacks**, exhibiting features of a clasic **plant succession** (see chapter 8). Dunes are threatened by human activity such as trampling and, therefore, have to be managed to prevent **blow-outs** which may destroy the dune system.

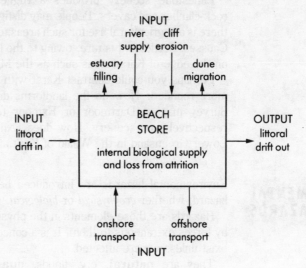

Fig. 6.14 The beach system

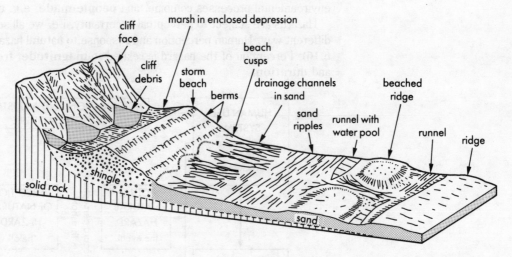

Fig. 6.15 Beach landforms

### Changes in sea level

Sea-level changes are a much longer-term phenomenon, resulting from the ice ages. Perhaps the **'greenhouse effect'** will result in rising levels over the coming decades/century. **Eustatic changes** are the changes brought about by ice-induced fluctuations. **Isostatic adjustments** are the movements of the crust resulting from the weight gain/loss from the presence of ice. The effects are **raised beaches, fossil cliff lines, rias, fjords, estuaries** and offshore shingle/sand-drowned beaches.

## GEOLOGY AND SCENERY: THE CASE OF KARST

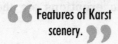

**Features of Karst scenery.**

Karst is the name of a region in West Yugoslavia which has given its name to the scenery associated with carboniferous limestone, although it is found in other younger limestone. **Chemical weathering, carbonation** and **fluvial erosion** dominate. The limestone has regular massive blocks and the joints influence the key features.

Water permeates via the joints or **grykes** between the blocks which form a **pavement** of **clints**. **Solution** takes place faster in the grykes. Running water widens these to form **swallow** or **sink holes** leading to **underground cave systems** whose location may be related to changing levels of the **water table. Blind valleys** occur upstream of a sink hole as downcutting continues. Caves may be **phraetic**, i.e. water-filled or **vadose**, i.e. containing an underground stream.

Surface depressions or **dolines** or **shake holes** are developed beneath drift due to intensified chemical weathering. The larger depressions or **poljes** are explained by both solution, faulting and water erosion when the water table was higher. **Gorges**, e.g. Cheddar, are close to sharp faults and were probably formed by retreating waterfalls rather than collapsed caverns. Gorges which contain rivers are well developed in the Causses region of France and are called **fluviokarst**.

Limestone scenery provides a rugged landscape attractive to tourists, walkers, rock-climber and cavers. People may disfigure areas by quarrying. Depending on location, there is an agricultural use for such areas, e.g. hill farming in the North Pennines and in the Causses. Settlement is rare, owing to the lack of water supply, and settlements tend to be on the edge of Karst areas such as the Mendips.

Perhaps you could contrast Karst with the scenery found in the chalklands. Why is it more rounded? How do the landforms differ? Similarly you could look at an Ordnance Survey map of Dartmoor or Exmoor to see the effects of granite and sandstone, respectively, on scenery. How does Exmoor differ from the sandstone scenery of the Lower Greensand in the Weald of Kent and Sussex?

## ENVIRONMENTAL HAZARDS

Environmental hazards are introduced here. The first part is relevant to all forms of hazard, whether *geophysical* or *biological*. Other hazards will be noted in chapters 7–9.

Hazards are those elements in the physical environment, harmful to people and caused by forces extraneous to them. It is a concept which is people-centred because it does not exist unless we are affected.

They are **natural**, e.g. floods, **quasi-natural**, e.g. smog, because people and environmental processes combine, and **people-made**, e.g. chemical spillages.

Hazards are subject to perceptual uncertainty, i.e. we all see or understand events in a different way. Human perception and response to natural hazards has been modelled (Fig. 6.16). Perception of the hazard is related to **magnitude, frequency, threat warning** and **duration**.

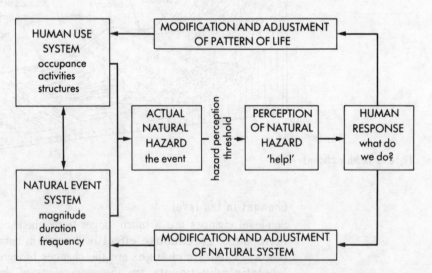

Fig. 6.16 Model of human perception and response to natural hazards (after R. W. Kates, 1971)

Hazards have a life-cycle with a **recovery time** after the event. People adjust to a hazard in a particular way as a result of various inputs to their decision-making processes (Fig. 6.17 and 6.18).

Geological and geomorphic hazards are one branch of geophysical hazards. **Avalanches, earthquakes, erosion, flood, landslides** and mass movement events, **river pollution, shifting sand, tsunami** and **volcanic eruptions** are the most common. (Some of these are covered in other sections.)

### Avalanche

Avalanche is the fast descent of a mass of ice, rock and snow from a mountainside where it is either partially thawed and unstable or recent and uncompacted. Their tracks may be mapped so that precautions may be taken but may follow unused paths and constitute a hazard.

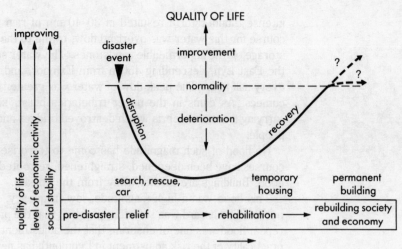

Fig. 6.17 Adjusting to a hazard event

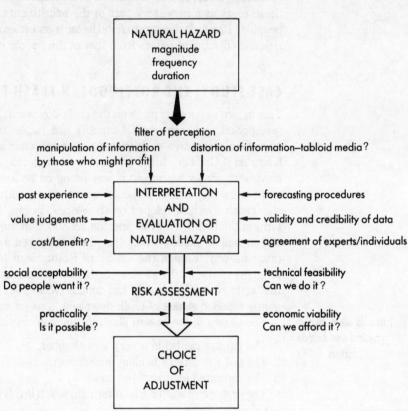

Fig. 6.18 Pattern of official adjustment to hazards

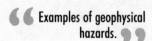

**❝ Examples of geophysical hazards. ❞**

### Mass movement

See under weathering and slopes. The Aberfan disaster 1966 resulted from a slide of coal-mining spoil triggered by springs beneath the soil heap.

### Shifting sand

Shifting sand may bury agricultural land or destroy an oasis. In the Netherlands dunes have covered fertile agricultural areas on the edge of Veluwe, but over a long timespan.

### River pollution

River pollution can come from a variety of sources: effluent from sewage works, or industrial spillages such as the spillage of toxic chemicals from the chemical works at Basle which killed wildlife down almost all of the Rhine in 1986. Also the movement of nitrate fertilisers in groundwater and run-off into streams may accentuate the dangers to people who drink the water. In 1988 aluminium was found in drinking water in Camborne, Cornwall, and was shown to be affecting health.

## CASE STUDY: FLOOD

August 1952 was very wet on Exmoor and in the first two weeks 20 mm of rain had fallen, keeping the water table high and the discharge of the River Lyn high. On 15 August an intense thunderstorm resulted in 30–40 mm of rain falling within a few hours. The only course for this water was overland flow, owing to the saturation of the soil and full surface storage on the impermeable sandstones. The water soon entered the steep-sided valley of the East Lyn, descending 450 m from Exmoor, and river fells were increasing by 30 cm every half hour. The peak flow of water was greater than the Thames at its peak, i.e. 520 cumecs. As dams in the upper tributaries burst, so fresh waves of water descended, carrying trees and debris which destroyed bridges and cottages in Lynmouth, drowning 34 people.

No flood of such magnitude has come to terrorise the village since then, but the river courses have been deepened, straightened, re-routed and widened to cope with a potential flood. Buildings are further away from the stream.

This flood was a high magnitude, low frequency, short duration event which shocked people. The hazard was sufficiently violent that the process of reaction to the potential of a repeat flood was one of ensuring that the adjustments were undertaken at high cost. The practicality of the risk adjustment in Lynmouth has never been tested, but people accepted these costs as a necessary part of the adjustments. Perhaps you could look at how the people of Bangladesh react to flood hazards and attempt to explain why their assessment of risk and adjustment differs from that of the people of Devon.

## CASE STUDY: THE ROTTINGDEAN FLASH FLOOD, 1987

This hazard event took place on the chalk downlands of Sussex. Several days of persistent rain (including 67 mm on 7 October) had raised the groundwater levels and reduced infiltration in this area of downland (see map) when an even heavier rainstorm fell at 1930 hours on 8 October. It fell on an area of moderate slopes which had been ploughed and sown with winter wheat, so it was falling on an area devoid of vegetation. The grassed areas were the steeper slopes. Run-off, mixed with soil and small chalk stones from the sheetwash which developed on the slopes, began to concentrate in rills which began to form in the fields. The rills and the flashflood in them joined in the valley flow to form a 140 cm wall of muddy water that then destroyed a barn and flooded 66 properties as it made its way through the village of Rottingdean to the coast. The only breaks in the progress of the mud wall were hedgerows, walls and raised road surfaces which ponded the water bank. The cost of the damage was estimated at up to £1 million. Here, on the gentle rounded slopes of chalk downland, a major hazard had struck.

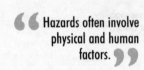

Hazards often involve physical and human factors.

The causes of the hazard illustrate the combination of physical and human factors:

1  The intense rainfall in a very wet October.
2  The soil left bare of binding grass cover.
3  The saturated nature of the soil.
4  The practice of winter wheat farming which involves preparing fields during wet autumn months.
5  The lack of barriers to the floodwater because of few hedgerows.
6  No stormwater drains large enough in the settlement.

The solutions are:

1  To do nothing and hope that it does not recur, but insurance companies will expect higher premiums in such a place.
2  To build earth dams up the valley – a costly exercise.
3  To alter the cropping patterns voluntarily (and lose money), or compulsorily (with compensation), or to remove subsidies for winter wheat growing.

Have you got a file showing the occurrence of such hazard events in your area?

# EXAMINATION QUESTIONS

### Question 1    Hydrology – a structured question
Study the hydrographs for two gauging stations (A and B) following a storm of 8 hours' duration (Fig. 6.19 a). The location of those gauging stations is shown on the map (Fig. 6.19 b).

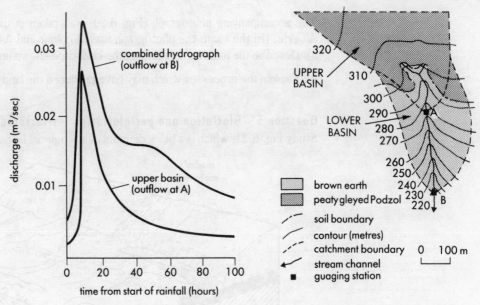

Fig. 6.19a)

Fig. 6.19b) *Source*: B. J. Knapp (ed.), *Practical Foundations of Physical Geography* (Allen and Unwin, 1981).

a)  i)  Explain what is meant by a *hydrograph*. (2)
    ii) Describe major characteristics of the hydrograph for Station A. (2)

b)  i)  Describe how the hydrograph for Station B differes from that for Station A. (2)
    ii) Suggest explanations for the differences you have described in (b)(i). (4)

c)  Explain how the hydrograph at Station B would be altered
    i)  if much of the drainage basin had been built over; (5)
    ii) if the drainage basin was underlain by a much more permeable geology. (5)

## Question 2    Plate tectonics – structured essay

a)  Define the following terms, which are all used in relation to the evolution of major land masses, and name ONE location for each:
    i)   shield area;
    ii)  sedimentary basin;
    iii) rift valley. (6)

b)  With reference to detailed examples, describe and explain the landforms, both above and below sea level, associated with
    i)  constructive plate margins.
    ii) destructive plate margins. (19)

(JMB 'B', June 1988)

## Question 3    Volcanoes – essay question

Describe and explain the global distribution of volcanoes. (25)

(AEB, June 1988)

## Question 4    Data response question

Fig. 6.20

The accompanying photograph (Fig. 6.20) was taken in the Dinosaur Provincial Park in Alberta. (In the exam the photograph was in colour and A4 size.)

a) Describe the main components of the landscape shown on the photograph.

b) Explain the processes which may have produced the landscape components.

(AEB, June 1988)

### Question 5   Glaciation and periglaciation – structured question

Study Fig. 6.21, which is a block diagram of a proglacial landscape.

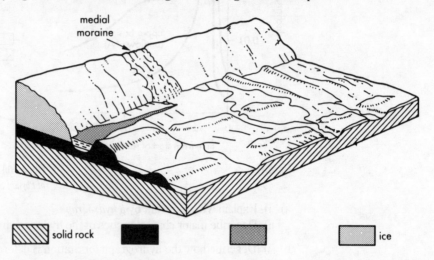

Fig. 6.21

| solid rock | | | ice |

a) Annotate the diagram by indicating a morainic ridge (M), an esker (E), and a proglacial lake (L). *(3)*

b) i) Complete the key to the block diagram by identifying TWO unnamed deposits. *(2)*
   ii) State THREE differences between the two deposits you have identified. *(6)*

c) i) What is a drumlin?
   ii) How might it have been formed?

d) Using the evidence from the diagram, suggest a sequence of FIVE stages accounting for the development of the landforms shown in the diagram.

(London)

### Question 6   Desert processes – essay question

Why are desert landscapes so varied?

(O and C)

*Note*   The trees in the photograph were planted
Fig. 6.22a)   about 11 years before the photograph was taken.   Fig. 6.22b)

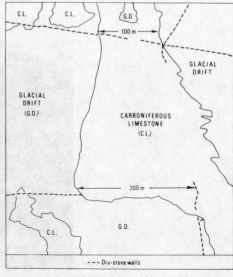

### Question 7 Karst – structured essay responding to data

a) What are the main landforms which you would expect to find in the upland areas of carboniferous limestone in Britain? *(10)*

b) The photograph and the geological sketch-map (Fig. 6.22a and b) are of an area of the Yorkshire Pennines. Study them carefully, then

i) name the landform features shown on the photograph; *(5)*

ii) discuss the past and present processes which have developed these landforms. *(10)*

(Oxford)

# OUTLINE ANSWER

### Answer to Question 3

The question has two parts, although it is not clear how you should divide up your time. As explanation is usually the more important, perhaps 70 per cent of your 45 minutes should be devoted to this.

1 *Description* – perhaps a map would assist, even if it is only of the Pacific 'Ring of Fire'. You could use this to stress the idea that there are chains of volcanoes and it is those chains which need explanation.

2 *Explanation* – should come from the ideas of plate tectonics. Volcanoes are concentrated along plate margins. Use diagrams to show their location relative to the destructive plate margins, both as land volcanoes and island arcs. Don't forget the 'hot spot' volcanoes, such as the Hawaiian Islands with their basic lavas. Finally, remember that volcanoes also occur along mid-oceanic ridges.

## STUDENT ANSWER WITH EXAMINER COMMENTS

### Question 1

" Poorly expressed. Discharge worth a little. "

" Confused. "

" Gets peak. "

" Yes. "

" Not descriptive. "

" Poor. "

" Run-off is not discharge. "

" No. "

" Only one point made. "

" Examiners will allow a slip once but not often. Discharge, not surcharge. "

" Don't make silly mistakes. "

" No idea of rapid run-off. "

" Again. "

a) i) Hydrograph is meant by the measurement of rainfall or water eg Discharge.

ii) The characteristics are there are less and the rainfall in relation to discharge goes down quickly. It peaks at 18 hours and rises to about 0.027.

b) i) Hydrograph B has a larger peak, over 0.03. It then very gradually declines after 100 hours to 0.01 discharge. Because it is at the bottom of the drainage basin so it gets more run off or discharge, which it gets only above and on top of it.

ii) The differences are because area that to measure its surcharge will be counted by B after the surcharge runs down the Basin. All the run off will join the surcharge that is measured by B. This is why it takes so long for it to go down after so much time, because B is at the bottom of the drainage basin.

c) i) Some surcharge would have been intercepted by building etc. Drainage would then divert some of it, so there would be less surcharge and it would fall rapidly after rainfall. The bank would have been levelled, so it would not matter about how high the water comes from.

ii) The surcharge would be reduced again as it would not measure underground run off. The water table would be increased. The graph would decline rapidly as most water would be taken into the soil.

" One point, what about base flow? "

" The answer is poorly expressed. Ideas are vaguely understood. Answer might just pass but it does not do what the question asks. Try not to say one point in two ways. "

CHAPTER

# 7

# ATMOSPHERE

CIRCULATORY
SYSTEMS

SECONDARY
CIRCULATORY
SYSTEMS

TERTIARY
CIRCULATORY
SYSTEMS

CLIMATIC CHANGE

MICRO-
CLIMATOLOGY

GLOBAL CLIMATIC
ISSUES

LOCAL CLIMATIC
HAZARDS

WEATHER AND
ECONOMIC ACTIVITY

## GETTING STARTED

Many of you tend to omit this section of your syllabus, either because it has not been taught or because you think it is too difficult. However, if you have studied this part of a syllabus it is well worth considering as a part of your revision programme. This is because it can, and does contain, easier (or at least more straightforward) questions than many other sections of the syllabus. There is increased scope for practical questions, owing to the variety of data sources ranging from weather records to meteorological maps and, nowadays, satellite images of weather systems. Weather and climate is now a very popular topic in the media, with the interest in the ozone layer, the 'greenhouse effect', acid rain, and the frequency of other hazards such as hurricanes. There is a danger here for the good candidate. Too much of the media interest is at the level of 'colour supplement' geography. You need to know the processes which lie behind the concerns of the moment.

**Solar radiation** is the basic driving force of the global atmospheric system. It either reaches the earth or is lost as a part of the **solar-energy cascade**. The radiation received is unevenly distributed and so it has to be redistributed either by **ocean currents** or from the tropics to the poles by the **atmosphere**.

# ESSENTIAL PRINCIPLES

**Circulation** is the movement of the atmosphere in response to **pressure, gradient force, Coriolis fine** and **frictional force**, which produce a diversity of airflows, namely **primary, secondary** and **tertiary circulatory systems**.

**CIRCULATORY SYSTEMS**

## PRIMARY CIRCULATORY SYSTEM

The basic model of the primary circulatory system is shown in Fig. 7.1. The system contains three **cells**. Between the cells at high altitude there exist **jet streams**, high-speed westerly winds (easterly in the tropics) which affect the movement of the cells and cause waves in the upper westerlies called **Rossby waves** which then control the medium-term location of the major pressure systems or cells.

Secondary systems are the main pressure cells, i.e. **cyclones** and **anticyclones**.

Terteriary systems are smaller in scale, such as land–sea breezes and valley winds.

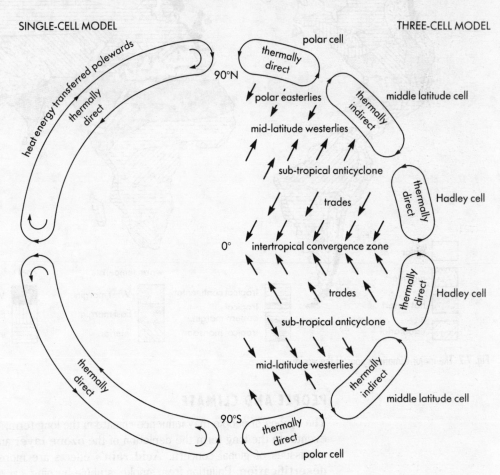

SINGLE-CELL MODEL

THREE-CELL MODEL

**Fig. 7.1** Models of the circulation of the atmosphere

## Water

Water, as we saw in chapter 6, moves through the **hydrological cycle**. **Evaporation** takes place from water bodies and **transpiration** from plants and, combined, they are called **evapo-transpiration**. Water **condenses** in the air at or below the **dew point** when the air is saturated. Water condenses around **hygroscopic nuclei** such as salt and dust, when air is often only 80 per cent saturated, i.e. **relative humidity** of 80 per cent. Condensation forms due to **contact cooling, advection cooling** and **adiabatic cooling**.

## CLIMATIC REGIONS

There are different climatic regions on the globe (Fig. 7.2). Each climatic region has a specific location and features of its weather pattern in terms of temperature, precipitation, hours of sunshine, humidity, wind direction, dominant pressure system or influential airmass. The diurnal and seasonal variations in the weather combine to make the climate of that place. Note which climates you should study and have basic examples of each type.

Secondary circulatory systems affect the pattern of weather most, and it is the changes which these bring to an area which most influence patterns of economic activity. The systems can constitute a hazard if their effects are exceptional. Thus a **temperate cyclone** or **depression** could almost reach hurricane force, as on 16 October 1988 in the UK. Tropical storms can become **hurricanes**. A dominating anticyclone over the UK in summer could result in drought, as in the summer of 1989. Extreme low winter temperatures result from a similarly dominating winter anticyclone.

Weather can be considered a hazard either as a result of secondary systems, e.g. **blizzard, fog** or **high winds**, or as a result of tertiary systems, e.g. **frost** from valley winds such as the **mistral**, or avalanches produced by the **Föhn**.

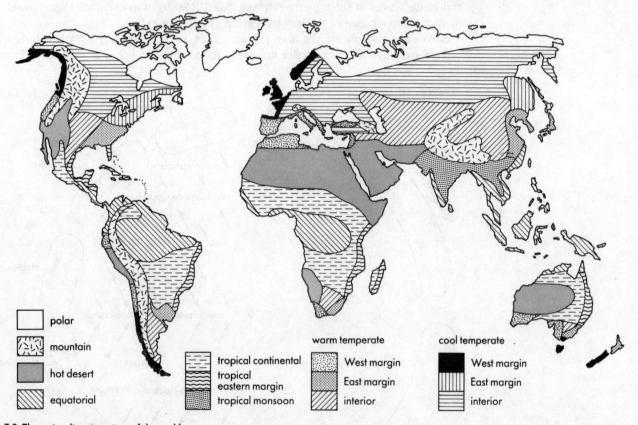

| polar |
| mountain |
| hot desert |
| equatorial |

tropical continental
tropical eastern margin
tropical monsoon

**warm temperate**
West margin
East margin
interior

**cool temperate**
West margin
East margin
interior

Fig. 7.2 The major climatic regions of the world

## PEOPLE AND CLIMATE

The actions of people may influence climate in the long term, or even short-term weather events. In the long term the depletion of the **ozone layer** and the 'greenhouse effect' are issues of global concern. **Acid rain**'s effects are more localised, as are those of **desertification**. Pollution from smoke, stubble-burning, car exhausts and other fossil fuel burning is of increasing concern. Urban areas may trigger off **convectional** storms and in winter produce **heat islands**. Conversely, climate and weather do influence economic opportunities.

### Micro-climates

Climates within a few metres of the ground in a small area, known as micro-climates, reveal many complex variations. Urban areas, forest, grassland, reservoirs and even a tree or windbreak all give rise to distinct variations in climatic variables at different times of the day and season.

## STABILITY AND INSTABILITY

Temperature falls with height: this is the **environmental lapse rate** (Fig. 7.3A) which varies according to time and place. If air rises its temperature will fall – the **adiabatic**

**lapse rate**. If that air is unsaturated it will cool at the fixed rate of 9.8°C per 1000 metres – the **dry adiabatic lapse rate**.

Once that air reaches dewpoint, or **condensation level** (Fig. 7.3B), condensation begins and, if it continues to rise, it will cool at a rate between 4°C and 9°C per 1000 metres – the **saturated adiabatic lapse rate** (Fig. 7.3C). It is lower because latent heat has been released through condensation. If the dry adiabatic lapse rate is to the left (see diagram) of the environmental lapse rate, the air is **stable**.

**Conditional instability** (Fig. 7.3C) occurs if rising air, cooling at the saturated adiabatic lapse rate, becomes warmer than the surrounding air and, at that height, it will rise freely and become **unstable** (Fig. 7.3D).

**Inversions** (Fig. 7.3E) occur when temperature rises with height owing to cooling of the lower airs of the atmosphere.

> " Causes of stability and instability in air. "

Fig. 7.3 Lapse rates, stability and instability

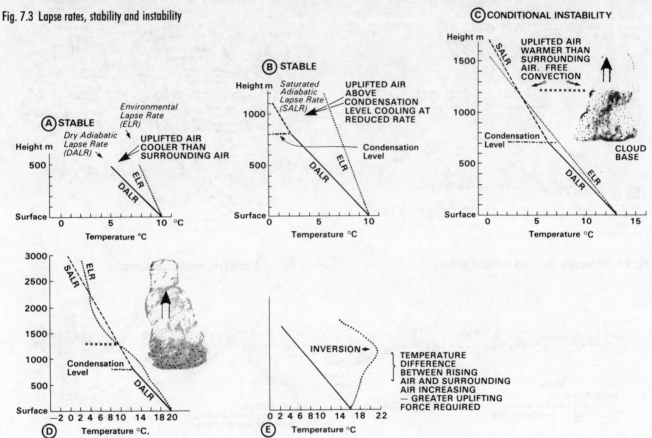

## THEORIES OF PRECIPITATION

- **Bergeron Theory** – ice crystals grow at the expense of supercooled water droplets. Crystals fracture and provide more *foci* for crystal growth. These crystals form snowflakes which fall, melt and become raindrops.

- **Collusion and Coalescence** – in rapidly rising warm clouds, droplets collide and fall through cloud, coalescing, breaking up and colliding again.

## SECONDARY SYSTEMS IN HIGH LATITUDES

**SECONDARY CIRCULATORY SYSTEMS**

### Anticyclones

Anticyclones are areas of high pressure. They dominate mid-latitude contental areas in winter, leading to intense cold in Central Europe and occasional cold spells as far west as Great Britain, i.e. a cP (continental polar) **air mass**. The associated hazards of frost, ice and snow on the east coast are associated with this high. The Azores high is an mT (maritime tropical) **air mass** which reaches Britain, often associated with the **warm sector** of depressions. High pressure over the UK in summer can be called a **blocking anticyclone**, because depressions are pushed north, and is caused by enlarged Rossby waves moving the jet-stream track and its associated depressions northwards. This was the cause of the 1976 drought in southern Britain.

## Depressions

Depressions are formed along the Rossby waves where **divergence** occurs in the upper air and permits **convergence**, i.e. low pressure at the surface (Fig. 7.4).

As a result, belts of rapid temperature change form, i.e. **fronts**. Depressions have been modelled as shown in Figs 7.5 and 7.6.

The weather map shown in Fig. 7.7 indicates the conditions at 0600 hours on 12 January 1978. Can you indicate the possible changes in the weather at different points on the map? Do you know what the symbols mean?

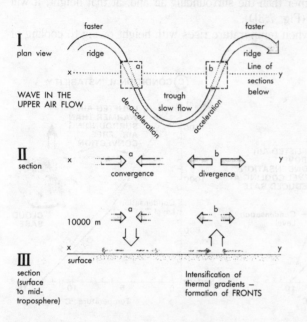

Fig. 7.4 Convergence, divergence and upper air flows

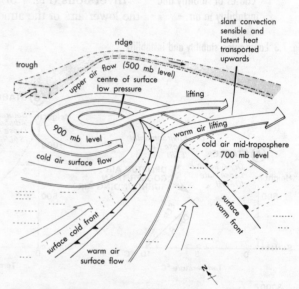

Fig. 7.5 A model of a mid-latitude depression

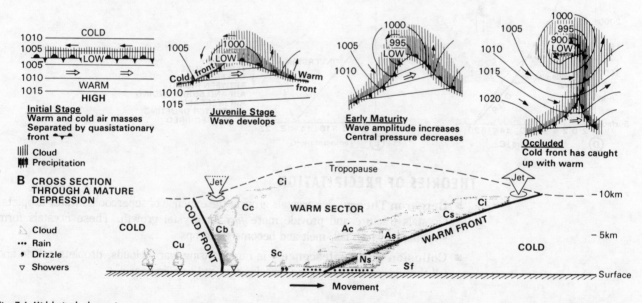

Fig. 7.6 Mid-latitude depressions

# SECONDARY SYSTEMS IN THE TROPICS

## Anticyclonic conditions

Anticyclonic conditions exist equator-wards of the sub-tropical jet and are associated with hot desert climates. With descending air, clear skies, high incoming radiation and low precipitation levels, the environment is a hostile one. The anticyclonic area fluctuates – in both medium and long term – and therefore the area of hot desert climate does advance and retreat. The Romans grew wheat in what is now the north Sahara.

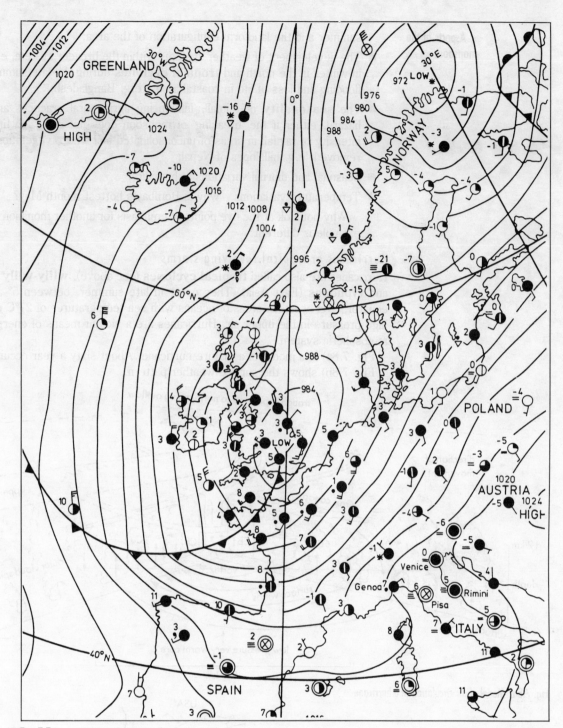

Fig. 7.7

**Desertification** is the extension of desert conditions into areas where they did not exist previously. The cause is partly climatic – the **inter-tropical convergence zone** (ITCZ) does not shift far enough to bring summer rains to the desert margins and savanna areas. The reasons for this may be related to other aspects of radiation receipt and ocean currents in the global energy cascade. However, people have played their part in accelerating the process in recent years, particularly in the Sahel area of Africa. Population growth, pressure to grow cash crops for export, both in colonial times and today, overgrazing and scrub clearance for timber for cooking have all contributed to the advance of the desert in Chad, Mali and Burkina Faso (see chapter 8).

### The monsoon

Monsoon climates are found in areas of reversed winter and summer air flows, e.g. South and East Asia, North Australia, East Africa and Guinea coast of West Africa.

Emphasis should be placed on:

- The marked seasonal shift in wind systems caused by the position of **jet streams** in the planetary system, the existence of an intense thermal low over the Indus valley in

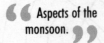

> **Aspects of the monsoon.**

summer and the landform configuration of the area;

- The development of weather systems within the broader system, e.g. **depressions** in winter in the north and **tropical cyclones** during the late summer often causing flooding and loss of life in coastal areas, e.g. Bangladesh;

- The **seasonality** of rainfall, influencing agricultural practices and constituting a hazard, either if the 'breaking' of monsoon does not occur on time or caused by intensity of rainfall in areas of unconsolidated soil where vegetation cover has been removed, e.g. hillslopes of Nepal;

- Seasonal and diurnal variations in humidity;

- Temperature variations – why is Bombay's hottest month May?

    Why not look at the five points of emphasis for another monsoon area so that your example is different?

### Hurricanes or tropical revolving storms

Hurricanes are also called **tropical cyclones** (see above), **willy-willy** (North Australia) and **typhoons** (East Asia). They occur in late summer, between 5° and 15° from the equator, over western areas of oceans with sea temperatures of 27°C or more and with high pressure in the upper air. Hurricanes are a violent means of energy transfer in the atmospheric system.

Fig. 7.8a) is a model of a mature hurricane. About sixty a year occur in total.

Fig. 7.8b) shows the surface weather pattern.

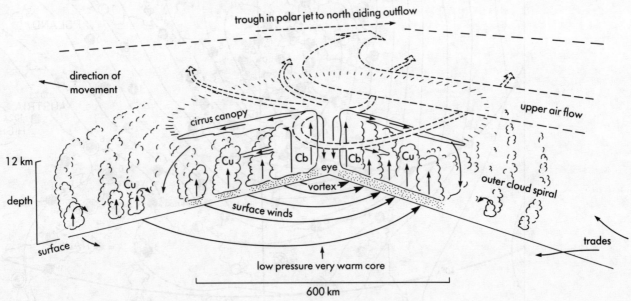

Fig. 7.8a) Model of air circulation in a hurricane

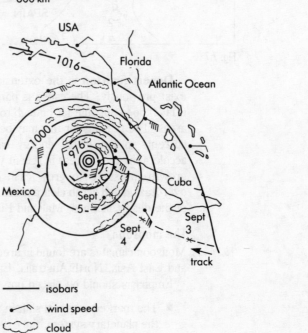

Fig. 7.8b) Surface weather map of a hurricane

The **eye** is adiabatically warming descending air, clear and calm. The surrounding **vortex** has towering cumulo-nimbus clouds with rapid uplift, torrential rains (500 mm in a day) and spirals due to low friction over the sea. Wind speeds here of over 160 km/hour give rise to heavy seas. They follow *tracks* which curve away from the equator, have a duration of 14–21 days over the sea, moving several hundred kilometres, and eventually dissipate over land where increased friction alters winds from spiral to centripetal so **filling** the low pressure.

Hurricanes are a major hazard, although perception of them varies (see chapter 6). Warnings help but the reactions to warnings vary. Contrast the acceptance of the inevitable that TV pictures show of the effect of cyclones on the delta dwellers of Bangladesh and Bengal with the panic evacuations seen on the Gulf Coast of the USA, where the insurance values make the losses seem greater. However, the total loss of a banana crop in St Lucia, or a fishing fleet smashed by waves, or the evacuation of tourists and cancellation of holidays, as occurred in Jamaica following Hurricanes Gilbert (1988) and Hugo (1989) may have caused greater hardship to people. Adjustment to loss of life, property, economic livelihood, transport and communications varies with its perceived importance and recovery time will also vary, especially if insurance money or aid can assist the recovery process.

## TERTIARY CIRCULATORY SYSTEMS

### Land–sea breezes

Land–sea breezes are a product of solar energy cascade on a diurnal basis – it is at a **mesascale**. Land heats faster than water, which needs five times more energy to raise its temperature by a comparable amount. Air over land expands and pressure falls, drawing in air from over the sea: temperature falls and humidity rises. (As I write this, in late March, a sea–land breeze is drawing in sea fog and obscuring my view!) The reverse happens at night when land cools more rapidly. It can happen along shores of lakes, e.g. Lake Michigan – here the lake shore is cooler by day and warmer by night in the summer, which is more desirable and so property is more highly valued: is this the reason for Hoyt's upper-class sector in Chicago running north along the Lakeshore Drive?

### Anabatic and katabatic winds

The anabatic wind is an upslope wind caused by the greater isolation received on valley sides, heating and causing rising air which draws air up from the valley. At night colder upslope air is drawn downslope, or sinks (katabatic wind).

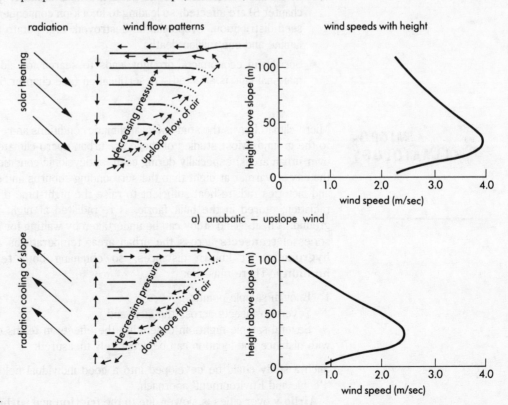

Fig. 7.9 Anabatic and katabatic airflow

a) anabatic – upslope wind

b) katabatic – downslope wind

**Föhn winds** (Alps-Austria), the **chinook** (Rocky Mountains), **Santa Ana** (California) and **Zonda** (Argentina) are dry warm winds on the leeward side of mountains. Rising air on the windward side is cooling at **saturated adiabatic lapse rate** as it rises over mountains and is warming at **dry adiabatic lapse rate** as it descends on the leeward side. Logically this should always be associated with precipitation on leeward side: the fact that this is not always the case has caused some to suggest that it is merely the forced descent of dry air from the upper atmosphere that warms.

The most rapid temperature rise of +21°C in four minutes has been recorded in Alberta. This accelerates the snow melt and causes the avalanche risk to increase. Avalanche protection measures and warnings are the norm in the Alps and Rockies.

## CLIMATIC CHANGE

Climatic change is a long-term occurrence and very often changes are possibly part of a cycle rather than a trend. There is a series of timescales over which change may be analysed. Over long timescales evidence can be gained from **ice dores, isotope analysis**, enabling temperature variations to be plotted, **geology** as a clue to past climates, **palynology** or **pollen analysis** – studying pollen grains in, for example, peat bogs – **dendrochronology** – studying the pattern of tree rings showing temperature and water variations for individual years – and **carbon dating** using carbon-14 dating.

Most recent changes may be apparent from weather records and observations which go back three centuries, and even from written accounts by (for example) diarists.

*" Causes of climatic change. "*

### Causes of change

1  The variations in energy output of the sun – eleven-year cycle of **sunspot activity**;
2  **Continental drift**;
3  Dust in the atmosphere from major volcanic eruptions (e.g. Mount St Helens in 1980) may reduce radiation for a few years;
4  Particles and gases from modern activities, e.g. $CO_2$ (see the section on the 'greenhouse effect' below);
5  Variations in patterns of circulation;
6  Short-term changes in interactions between atmosphere and oceans, e.g. **El Nino effect** – heavy precipitation in Peru, anomalous weather over Pacific Ocean and droughts in Africa.

### Effects of change

■ Biological – alterations in the pattern of plants and crops; parts of food chain (see chapter 8) are affected, so leading to knock-on consequences; wind changes will alter seed distribution. Crops can be destroyed, e.g. potato blight caused the 1848 Irish famine and one million deaths.

■ Social and economic – drought leads to search for aid and for new seed strains; marginal land is affected – desertification (see chapter 6).

## MICRO-CLIMATOLOGY

Micro-climatology is the study of local climatic conditions and especially those areas closest to the ground. Most studied of these are urban micro-climates. **Heat islands** are found over urban areas, especially during calm, anticyclonic conditions in winter. The centre of the city is warmer at night than the surrounding suburbs and countryside. Homes, offices and factories radiate heat sufficient to raise the night-time temperatures. Incoming solar radiation, stored in the built fabric, is re-radiated at night faster than from vegetated ground. A heat-island study can be undertaken by waiting for still winter nights when, in a series of **transects** across the urban area, temperatures are taken with a **whirling hygrometer** at known distances, so obtaining both **temperature** and **relative humidity**. There must be:

1  Enough sample points;
2  Several transects across the city; and
3  Several sample nights in order that the effects in terms of variations in temperature with distance and location can be accurately measured.

Such a study could be developed into a good individual field study approach within the 'People and Environment' approach.

**Airflow** over cities is slower due to the **friction** and **turbulence** caused by buildings, but buildings might act as wind funnels, so giving the impression of stronger winds. The effect of buildings in summer can increase the instability in the atmosphere owing to greater radiation of heat: so much so, that convectional storms build up and can result in heavier precipitation to the leeward side of the city.

Sunlight can be obscured by **pollution** over cities.

Microclimates may be studied in areas of different vegetation or crops. The micro-climate conditions in a deciduous woodland will differ from coniferous woodland. **Windbreaks** are designed to alter the micro-climate of the area downwind. These effects also make for stimulating project work of a detailed kind involving rigorous measurements and recording over a period of time.

## GLOBAL CLIMATIC ISSUES

## THE OZONE LAYER

**Ozone** is most concentrated 20–30 km above earth in the upper atmosphere, and filters ultra-violet light. The ozone layer is disturbed by **chlorofluorocarbon gases (CFCs)** because ultra-violet light causes the chlorine to destroy ozone molecules (1 chlorine molecule can destroy up to 100 000 ozone molecules). CFCs are used in aerosols – 62 per cent of the 800 million produced in the UK used them in 1988 – in plastic foams, refrigerants, and in fire extinguishers (similar gases called halons).

*Here we look at a number of global climatic issues.*

### Evidence

- The destruction is most 'visible' over Antarctica (see below) when 50 per cent of the layer was destroyed in 1987 to produce 'the hole';
- NASA notes a decline of 3 per cent per annum in ozone over the northern hemisphere, which is four times faster than previous predictions;
- Slight evidence for a 'hole' over Spitzbergen.

### Effects

- Ultra-violet radiation increasing – destroys marine and terrestrial **food chains** (see chapter 8) – could affect fishing and agriculture;
- Increase in skin cancers due to higher ultra-violet doses;
- Implicated in the '**greenhouse effect**' (see following section) because CFCs also limit the heat leaving the atmosphere.

### Reactions

- The UN Environment Programme (1980) first called for reduction; the Montreal Protocol (1987) called for a 35 per cent cut by 1999;
- Toleration of continued use of CFCs in Developing World until they can afford to change: this is a problem, because greatest potential growth is here;
- Increased control over use as a reaction to latest evidence – 1989 London Conference and Hague Conference – speeding-up changes.

## THE GREENHOUSE EFFECT

This is the popular term given to the consequences of increased carbon dioxide levels in the atmosphere: since 1958 they have risen by 10 per cent. $CO_2$ permits incoming solar radiation but traps outgoing **long-wave radiation** and this warms the earth.

### Causes

- Burning of fossil fuels, e.g. in 1950 industry used 1620 million tonnes: in 1980 the figure had risen to 5170 million tonnes;
- The clearance of tropical rain forest has reduced photosynthesis (see chapter 8) leaving higher levels of $CO_2$ in the air. The result is known as the **greenhouse effect**, i.e. a warming of the atmosphere. The effect is also produced by releasing carbon from burning fossil fuel.

### Evidence

- The warmest three years since 1961 were all in the 1980s;
- Six of the warmest ten years have occurred since 1970 (Fig 7.10).

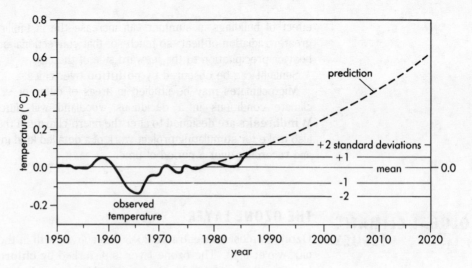

Fig. 7.10 Prediction of temperature increases from carbon dioxide warming

Fig. 7.11 looks at some of the impacts of changes in carbon dioxide on the climate, ecology, food production and society as a whole.

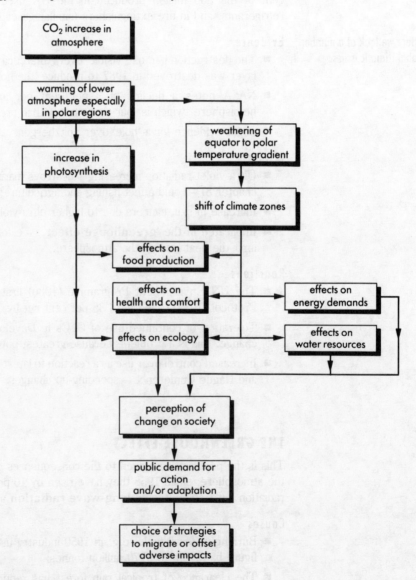

Fig. 7.11 Some of the interactions between carbon dioxide changes, climate, ecology, food production and society

### Reactions

- Take action to prevent or slow down change – the West German government is insisting on catalytic converters on car exhausts to cut emissions and on controls on power stations;

- Accept the inevitability and live with it or rely on market forces to bring about change – the 1989 British government is doing the latter and not demanding that privatised electricity take measures to cut $CO_2$ emissions.

## ACID RAIN

Acid rain or acid deposition refers to the people-made increase in acid brought about by air pollution. It is found in southern Scandinavia, North-West Europe, Eastern Canada, and North-East USA, although some of these areas are not the sources, e.g. Norway and Sweden. It is **trans-frontier pollution**.

### What is the evidence?

- **Forest dieback** (Waldsterben) experienced in West Germany and Sweden at first;
- Poisoned fish in streams and lakes.
- See Fig. 7.12.

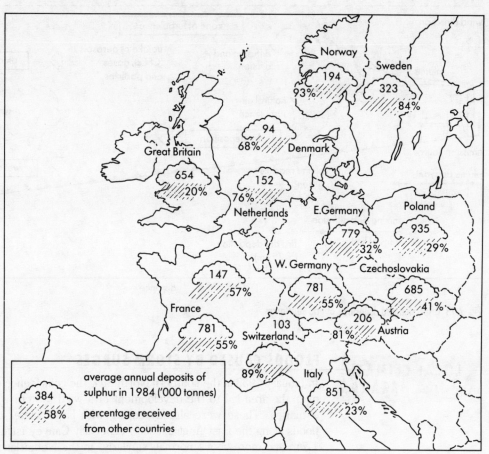

**Fig. 7.12 Where the sulphur falls**     Source: UN Economic Commission for Europe - EMEP Programme 1983

### What are the sources?

Burning of fossil fuels releases sulphur dioxide ($SO_2$) and nitrogen oxides ($NO_x$) which react with sunlight and ozone, and dissolve to produce acid rain. (In UK 70 per cent of acid rain is caused by sulphur products and 30 per cent by nitrogen.) Power stations (61 per cent of $SO_2$ in UK), domestic central heating and car exhausts are the main sources.

### How is it moved?

By weather systems.

### How is it deposited?

Fig. 7.13 (p. 90) looks at the dispersion and deposition of acid substances.

- **Dry deposition** close to sources, damaging stonework of, for example, old churches;
- **Wet deposition** in precipitation downwind of source. It may react with other heavy metals in soil to increase poison levels in soil – **indirect acid effects**.

### What are the solutions?

- Reduce sulphur emissions during burning – uses vast quantities of limestone – what do you do with the gypsum produced?
- Reduce sulphur from waste gases after burning in chimneys – again, uses limestone;

- Take sulphur from coal prior to burning – makes coal expensive;
- Switch to low sulphur fuels;
- Burn less fossil fuel – use nuclear power (1989 UK government favours this option), water and geothermal power where available. All cost money and will push up charges (£1 500 million in UK alone).

### What have countries done?

The 30 per cent club: in 1983 a group of 40 countries agreed on a reduction of emissions by 30 per cent by 1990s. (UK and USA have not signed.)

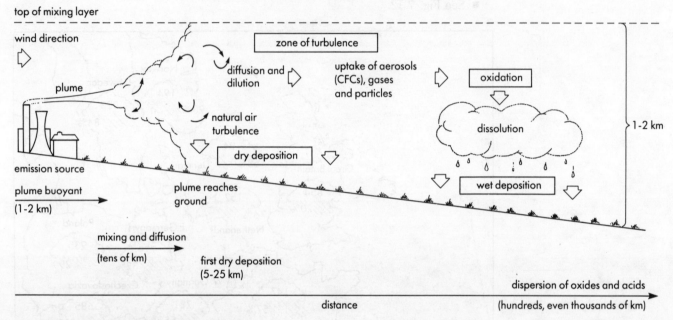

Fig. 7.13 Dispersion and deposition of acid substances

## LOCAL CLIMATIC HAZARDS

### FLOODS CAUSED BY STORM SURGES

In January 1953 in the southern North Sea, the coincidence of high tides with a depression over the area led to the development of a wind system which funnelled the tide up the estuaries. These were already coping with large winter discharges. The results were floods along the East Anglian coast, especially Canvey Island where levels were recorded 1.7 m and more above normal; similarly, in the Netherlands and West Germany. In the Thames estuary it was 2.75 m above normal high tide, and the flooding killed 300 people, with some 30 000 more being evacuated. The reaction was to invest in higher sea defences (raised by 2 m around Hamburg) and to build the Thames barrier (1983) at Woolwich to protect the 116 sq km of London which a repeat surge might affect. Osaka has *five* barriers to prevent surges. There is now a storm tide warning system. A similar surge can occur in the Bay of Bengal during a tropical cyclone: in 1970 200 000 were killed in one surge – what measures can be taken to reduce its effects there?

### TORNADOES

Tornadoes are violent spiralling vortices of rising air. They are associated with **squall-line thunderstorms**, i.e. storms associated with a cold front. They are most noted in Great Plains of the USA, known as 'Tornado Alley', in spring and early summer (66 per cent April–June). A **funnel** develops from a thundercloud where a very unstable **lapse rate** has developed, aided by **jet-stream** effects. The funnel is up to 150 m wide at ground level and will travel up to 10 km, its **explosive convection** and low pressure destroying all in its 160-mph path. In 1974 a total of 307 people were killed by one tornado; property damage averages $50 million per annum. Over sea they form **waterspouts**. Less violent forms do occur in Britain.

**Dust devils** are desert whirlwinds, although they are not linked to thunderstorms – they are caused by very localised instability in very calm, stable air.

Is it possible to warn people of tornado risk?

## DROUGHT

Drought is a lack of water in relation to its expected amount. In 1976 a **blocking anticyclone**, extending as a ridge of high pressure, resulted in a lack of frontal rain over southern Britain until the slowly shifting jet stream pulled the depression track south in the autumn.

Drought can be aided by human activity, as occurred in the 'Dust Bowl' in the mid-west USA in the 1930s. Drought will affect agricultural output. In July 1988 much of the US maize crop withered due to prolonged drought. It might affect water consumption if associated with high temperatures, and so exacerbate the problems of water supply caused by lack of water to replenish reservoirs and groundwater supplies.

Some areas where drought may affect livelihoods have taken unusual steps, e.g. Jersey has a desalinisation plant to enable 30 per cent of the demand for water supplies to be met from seawater. In Guernsey water is metered, and during the 1976 drought it was restricted to 1 cubic metre per household per week.

## FOG

Fog is produced when **condensation** at or near ground level causes visibility to fall below 1 km. **Advection fog** is formed when moist, warm stable air crosses a cooler surface and the lower layers cool, inverting the temperature; once dewpoint is reached, condensation occurs, e.g. Grand Banks off Newfoundland, where it was a hazard to shipping until the advent of radar. **Steam fog** is a form of advection fog: cold air passes over warm water which appears to steam as condensation takes place.

**Radiation fog** is formed by nocturnal cooling and loss of heat enabling dewpoint and condensation to occur. It often forms in valleys and hollows but can be trapped by a **temperature inversion**. It is a hazard especially to fast-moving road traffic, e.g. on the M1 near Newport Pagnell. The cost is traffic disruption and in many cases loss of life and destruction of vehicles: i.e. insurance cost. **Smog** or **photochemical fog** is used to describe radiation fog over urban, industrial areas when smoke, $SO_2$ and $CO_2$ are trapped beneath a temperature inversion. These may develop under anticyclonic conditions in summer or winter.

In 1952 an estimated 4000 people died from smog-related diseases in London. Over four days visibility was down to a few metres, and a brown-yellow coloured smog left a film of black even inside homes. Films and the opera were cancelled. The reaction was the introduction of the Clean Air Act in 1956 to curb emissions of at least the visible pollutants.

Los Angeles residents in the 1950s noted crop damage and eye irritation which was eventually found to be caused by the reactions in sunlight between the exhaust gases of vehicles and ultra-violet light to produce ozone which can induce headaches, nausea, and injury to the lungs, as well as impotency. As a result, in 1960 California passed laws to control vehicle emissions. The USA followed in 1963. The catalytic converter introduced in 1975 is beginning to reduce further the other harmful emissions from vehicles.

**Frontal fog** is associated with the passage of a warm front and saturation of the cool layers of air near the ground in advance of the front.

## LEAD POLLUTION

In 1983 the Royal Commission on Environmental Pollution recognised the danger of lead poisoning from, particularly, the addition of lead to petrol. In 1988 and 1989 taxation on unleaded petrol was made cheaper so as to discourage the use of leaded petrol in cars. Roadside soil will contain up to 0.3 per cent lead deposited from car exhausts. Lead results in mental damage, especially to children, and so efforts are being made to reduce atmospheric lead pollution which will build up during calm radiation fog-inducing conditions.

## FROST, SNOW, BLIZZARD AND HAIL

### Frost

Frost occurs when the air temperature in a **Stephenson's screen**, i.e. 1 metre above ground, falls below freezing point. Frozen dew may be called **white frost**.

**Black ice** is when condensation onto a frozen road surface itself freezes. It is a major hazard to traffic as its occurrence is invisible to the speeding motorist.

**Freezing rain**, i.e. rain falling onto frozen surfaces and becoming an icy film, has similar consequences. In southern UK the shoe-electric trains have also been affected by ice on the rail preventing electricity reaching the motors. De-icing trains normally prevent this hazard but are a heavy cost.

### Snow

On average, snow requires a freezing level of 300 m or lower to produce 23 mm of precipitation. Its occurrence is a hazard for traffic movement and often for agriculture, especially in upland areas.

### Blizzards

Blizzards are snowfalls associated with storm conditions. In Buffalo, USA, in January 1977 there was already 1 m of snow after forty days with snowfall. A blizzard with 70 mph winds struck, and in four hours all transport had stopped. Over 17000 people were trapped in offices, and some were rescued from cars up to 9 m beneath the drifts. It took two weeks to clear and 29 people died, nine frozen in their cars. Many rural families were isolated for ten days. The cost of clearing the snow, needing 1000 people, as well as the damage was estimated to be $250 million. Some of the cleared snow did not melt until May.

### Hail

Hail or ice stones with a diameter of 5–30 mm is associated with rapid convection, especially in **cumulo-nimbus** thunderclouds. Where violent hailstorms are common, e.g. mid-west USA, crop damage can cost $1 billion a year. Fruit is particularly vulnerable. The impact of this hazard, like tornadoes, is difficult to forecast, as its path across a landscape may be only a kilometre wide.

All these hazards should be considered from the viewpoints outlined in chapter 6.

| WEATHER AND ECONOMIC ACTIVITY OR ECONOCLIMATE |
|---|

Atmosphere is a source which can bring benefits, e.g. nitrogen from the air can be fixed by plants like peas. Weather and climate affect patterns of tourism and the type of facilities in resorts in areas of almost guaranteed sunshine, e.g. Barbados, compared with the British seaside resort. Fine weather will result in heavier traffic and jams on routes away from popular resorts. Winter sports are similarly affected by weather: centres with less guaranteed snow often have more alternative activities, e.g. Aviemore.

Retailing is affected by climate: a wet summer will often see many 'mid-season' sales to help dispose of stock.

Motorway building was affected in the 1984–85 winter when abnormally cold spells prevented concrete from being laid. Building work in general is affected by poor winter conditions.

Weather affects the yield of crops and, even in the UK, spray irrigation is used to provide consistent rainfall conditions for crops. Hay needs good drying weather; ploughing demands drier periods and crop spraying requires still, calm conditions.

# EXAMINATION QUESTIONS

### Question 1

With the aid of appropriate examples describe and account for the ways in which an urban area in temperate latitudes may modify the climate over and around the city. *(25)*

(AEB, June 1985)

### Question 2

Examine the main factors which control the direction and velocity of surface winds.

(London 210, Jan 1985)

### Question 3

a) Account for the aridity of the coastal region of Peru. *(5)*

b) Describe and account for the agricultural production of the coastal region of Peru. *(10)*

c) i) Describe the changes that have taken place in the organisation of agriculture in the Andean region of Peru in the last 20 years.

ii) Discuss the effect of these changes on agricultural output.    *(10)*

(JMB 'C', 1988 Alternative C)

This question begins with climate and asks you to show its influence on agriculture and possibly on agricultural change in the Andes.

## Question 4

Explain the background to the increasing concern for the impact of human activity on atmospheric processes. What are the reasons for this concern?    *(25)*

(AEB, June 1989)

## Question 5

The graphs below (Fig. 7.14) illustrate the climate at three stations – Kandi, Rangoon and Singapore, which are located on the accompanying map (Fig. 7.15).

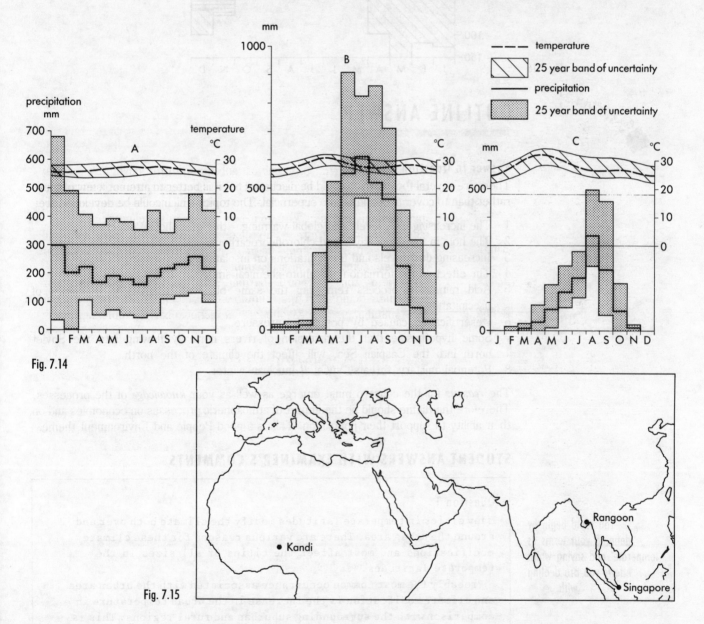

Fig. 7.14

Fig. 7.15

a) State the correct location beneath each graph.    *(3)*

b) Explain why the temperature regime is more stable than the precipitation regime in all three cases.    *(3)*

c) Complete the key to the diagram overleaf (Fig. 7.16) which suggests the effects of the rainfall regime at B on the soil moisture balance.    *(4)*

d) What is the effect of the rainfall regime at B and the associated soil moisture balance on agricultural practices in the region?    *(5)*

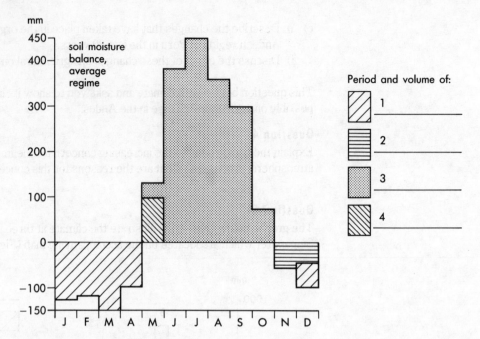

Fig. 7.16

# OUTLINE ANSWER

### Answer to Question 4

There are several topics which could be discussed but it is better to attempt a few of these rather than to cover a lot and be too superficial. The topics which could be developed are:

1   The increasing $CO_2$ levels and global warming – the 'greenhouse effect'.
2   The hole in the ozone layer and chlorofluorocarbons.
3   Increasing dust levels and the limitations on insolation.
4   The effect on fog formation and photo-chemical smog.
5   Acid rain as a process remaining the same but with changes in outcomes of precipitation.
6   Desertification caused by population pressure.
7   Some hypothesise that the diversion of rivers normally flowing into the Soviet north into the Caspian Sea, will affect the climate of the north.
8   Potential military threats, such as nuclear winter.

The *reasons* for the concern must emerge as well as your *knowledge* of the processes. Therefore the theme should be the impact of atmospheric processes on economies and on their ability to support their population. This is a good People and Environment theme.

## STUDENT ANSWERS WITH EXAMINER'S COMMENTS

 You could begin by defining your terms as 'temperate' and saying which effects you are dealing with.

### Question 1

Many cities in temperate latitudes modify the climate both over and around the urban area. There are various reasons for these climate modifications and most affect the cities of all sizes in the temperate latitudes.

Probably the most common occurrence associated with the urban area and climate modification is the increase in the urban temperature in comparison with the surrounding suburban and rural regions. This is most predominant in large built up cities such as London with its large tower blocks and closeness of the buildings. The following diagram shows the influence of the city of London over the temperatures in the area.

This diagram shows that, in mid-May, the temperature of London city centre is up to 3°C greater than that of its surrounding suburban and semi-rural area. Although this is not a great mathematical difference, climatically this increase has serious affects.

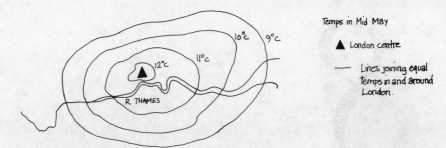

Temps in Mid May

▲  London centre

—  Lines joining equal temps in and around London.

Primarily, there is the effect on the amount of rainfall reaching the surface. Because of the number of buildings, interception is high but with the added increase in temperature, there is a marked increase in evapotranspiration. This could be a disadvantage in cities that often rely on rainfall for the washing of monuments and other public attractions, but could be a major advantage as much rain affected by carbon monoxide from exhaust fumes of motor vehicles will not reach the surface in as great amounts.

> The point about washing is irrelevant and suggests that you are not sure of your material. The link to fog and smog is incorrectly explained, although the effects are correctly noted.

The increase in temperature also has other consequences as severe changes in weather conditions often result in city smog. This causes severe problems to motorists and the city public as often the smog contains poisonous gases such as carbon monoxide and sulphur dioxide. This is the problem in Los Angeles where vast areas of the city area are affected by smog and fog due to the increased temperature of the city area and to the input into the air of motor vehicle fumes and industrial outputs such as sulphur dioxide.

There are many reasons for the modification of the urban climate which causes such occurrances as temperature increases, fog and increas evapotranspiration. With increased technology in architecture there has been significant effects on urban climates. Firstly, there is the better insulation of buildings which makes the buildings warmer so increasing the ground or building level temperature. This is true in many modern cities such as New York where urban temperatures have increased significantly with increased technology.

Because the buildings are warmer, there is greater heat escape from air conditioning units which expel warm air from buildings into the surround air above and around the city. The cause significantly less rain to reach the ground as often it is evaporated before reaching the surface. This reduces city run off and increased building evapotranspiration. Motor vehicles in the city have also caused climatic modifications mainly because of the heat output of their engines and the output of gases into the atmosphere.

> By focusing mainly on temperature you ignore wind, convectional rainfall and, particularly, the effects downwind of the city. This essay would possibly obtain a bare pass.

Therefore, by using examples such as London, New York and Los Angeles, it is clear that urban climates have been modified by the urban area as well as affecting the surrounding area and the air above the city. Such causes as air conditioning, central heating systems in buildings and motor vehicles have affected these climates in the cities of temperate latitudes.

## Question 5

> Correct answers.

a)   A SINGAPORE       B RANGOON       C KANDI

b)   All three stations are located in tropical sites near the equator and so temperature is stable since the sun is always overhead. However cycles of wind movements and pressure system variations cause variation in rainfall.

> Vague response. Sun is not permanently overhead in this location. Make reference to the movement of the ITCZ.

c)   1. Deficit      2.      3. Surplus      4.

> Correct, so far as it goes. Incomplete.

d)   The soil moisture surplus in the summer months and the deficit in winter could mean that the agricultural activity is limited to the summer months, or that irrigation is necessary to provide a growing season all year round. The large soil moisture content in summer probably means that rice is the main local crop.

> The late recovery in (d) was not sufficient to raise the mark above a C grade for this question.

# ECOSYSTEM BIOGEOGRAPHY

## GETTING STARTED

Biogeography and ecosystems are an optional part of some syllabuses, and others do not state the content very explicitly. If this area is on your syllabus or your revision plans then there are certain basic principles which are common to all ecosystems.

Ecosystem modifications as a result of human activities is another common theme. Here you could have your own case study of modifications together with your opinions on the outcomes.

In many syllabuses biogeography is rightly covered alongside soils, and so you may need to ensure that you understand the ecological links between the **biotic** (this chapter) and the **edaphic** (chapter 9 – Soils).

# ESSENTIAL PRINCIPLES

## Basic themes in ecosystem

- **Ecology** is the study of interactions between plants, animals and their environment, and an ecosystem is all the living organisms in an area and its non-living environment. An ecosystem can range in scale from a fish tank to an ocean, a lawn to the prairie grasslands. There are basic principles to the operation and functioning of ecosystems in the **biosphere**. You need to know both the principles and how these can be applied to different ecosystems.

- Plants are normally found in combinations in a **habitat** known as a **community** or **association**. The plants within a community interact by **competing** for space, being **complementary** and by **depending** on the presence of others. Within a habitat changes take place which often involve a **succession** of plants until a **climax vegetation** is reached. This is normally in a state of **dynamic equilibrium**.

- People have had a major impact on the biosphere. We have **domesticated** animals, changed the genetic structure, e.g. IR8 miracle rice, to make rice more productive, and by cultivation we have **simplified** communities. We alter habitats by **afforestation, deforestation, hunting, irrigation** and by **ploughing** and, in marine circumstances, by building barrages or draining marshes. We also attempt to control our impact by conservation, preservation and protection of habitats.

- People's impact on the biosphere may not just be local: there may be global implications of deforestation, or more local effects of pollutants from, for example, waste disposal.

Here we look at a number of key concepts in biogeography.

## ECOSYSTEM AND ENERGY

Energy from the sun enters the ecosystem via **photosynthesis** which occurs on **primary producers** to produce plant materials or **biomass**. The material or **net production** of the primary producers is the first **trophic level**, or the first stage in the **food chain**. These are **autotrophs** which use $CO_2$, plus radiant energy (**phototrophic**) and oxidised inorganic elements (**chemotrophic**) for energy.

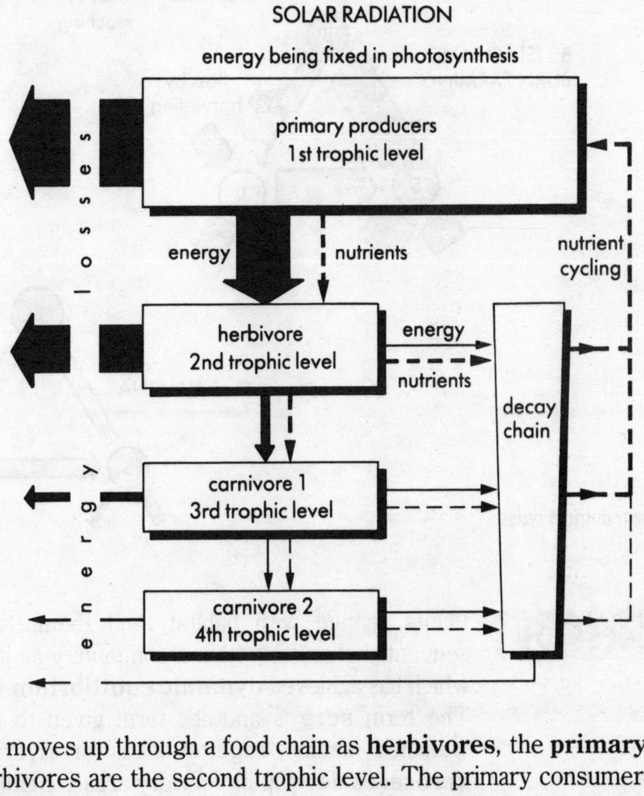

SOLAR RADIATION

energy being fixed in photosynthesis

primary producers
1st trophic level

losses

energy          nutrients

nutrient cycling

herbivore
2nd trophic level

energy

nutrients

decay chain

energy

carnivore 1
3rd trophic level

carnivore 2
4th trophic level

Fig. 8.1 An ecosystem: Energy flow and material cycling

Energy then moves up through a food chain as **herbivores**, the **primary consumers**, eat plants. Herbivores are the second trophic level. The primary consumers are eaten by **secondary consumers** or **carnivores** at the third trophic level and so on until perhaps people consume the organism. These are **heterotrophs**, i.e. they must gain energy from living or dead organisms.

At each trophic level less energy becomes biomass and energy is lost to heat. Because each level absorbs only 10 per cent of the energy received at the previous level, there are normally only four or five levels. Also, plants and animals at each level die and are decomposed by **decomposer organisms**, bacteria, fauna and flora. Food chains are normally more complicated **food webs**. Energy is almost continuously renewable.

**NUTRIENT CYCLING**

**Nutrient** or **material cycling** in the ecosystem is a **biochemical cycle**, but nutrients are *not* continuously renewable. Nutrients come from:

a) atmosphere, e.g. nitrogen;
b) the breakdown of rocks in the soil; and
c) decay. Some nutrients are returned by decay but generally human activity removes nutrients from the cycle. Decomposers rely on oxygen to return nutrients to the system.

There are five major sub-cycles: **carbon, phosphorus, nitrogen, sulphur** and **oxygen**. Nutrients are **stored** in:

> *Storage of nutrients.*

a) the **biomass** (most here in tropical rain forest);
b) **soil** (e.g. prairie grasslands); and
c) the **litter** (e.g. coniferous forest).

This can be shown diagrammatically for all systems.

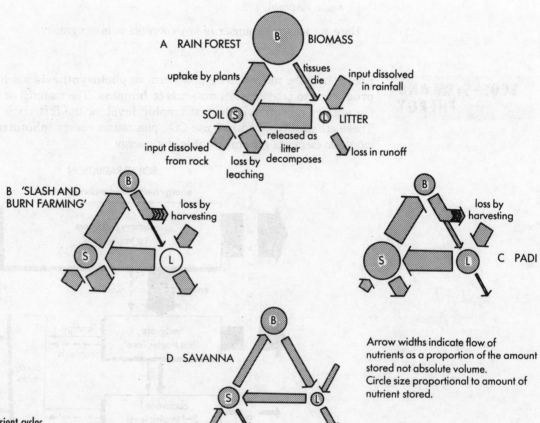

Fig. 8.2 Ecosystem nutrient cycles

**SUCCESSION**

Plants change with habitat, and themselves alter their habitat. **Succession** is the sequential changes within a community as it develops towards a **climax** or a community which has achieved **dynamic equilibrium** with its environment and especially its climate. The term **sere** is another term given to this development: **hydroseres** are found in wetlands on the edge of lakes or rivers, **psammoseres** develop in sand dunes, **xeroseres** develop in very dry areas, **lithoseres** are a sub-type of xeroseres developed on bare rock or scree, and **haloseres** develop in saline conditions.

## CASE STUDY OF SUCCESSION: THE SPRUCE BOG, ALGONQUIN PARK

The bogs in Algonquin Park, Canada, start in either glacial or beaver-created lakes. Near the edges a mat of sedges, sphagnum moss and shrubs is created which, as it dies, sinks to decay and form peat which gradually grows to meet the mat. The mat is invaded by spruce trees: the black spruce, which thrives in areas where the nutrient content is low, and later other forms of spruce. Gradually the advancing mat will fill the lake, sometimes aided by beavers felling the spruce until the climax vegetation of boreal forest is achieved. The process is a slow one and in some of the cases in the Algonquin Park it has taken up to 7000 years for the lake to disappear into climax forest (Fig. 8.3a)–d).

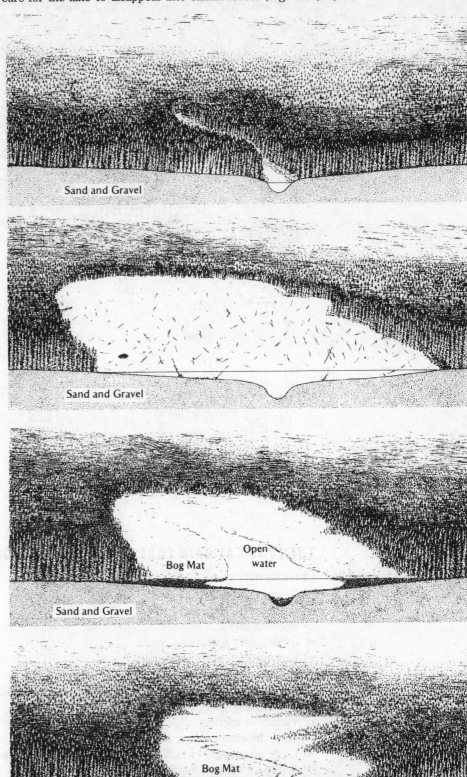

Fig. 8.3a) Cross-section of Sunday Creek as it was 8000 years ago – a stream through the forest

Fig. 8.3b) Cross-section of Sunday Creek as it was after beavers dammed it 7400 years ago

Fig. 8.3c) Cross-section showing how the floating mats advanced from each side towards the centre of the flooded valley

Fig. 8.3d) Cross-section of Sunday Creek today: Peat has filled the space below the two mats

## Subseres

**Subseres** or **sub-climaxes** occur when the development is stopped by topographic, edaphic (soil) or biotic factors. **Plagioclimax** occurs if people have arrested the succession, e.g. heather moorland in the UK, and even savanna grasslands.

### BIOMES

**Biome** is the term given to a major climax community. Fig. 8.4 shows the ten major biomes of the world, excluding mountain (montane) communities.

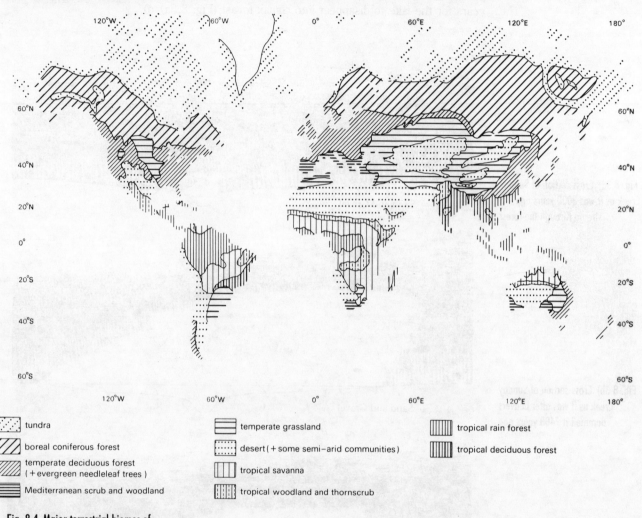

| | |
|---|---|
| ⣿ tundra | |
| ▨ boreal coniferous forest | |
| ▨ temperate deciduous forest (+evergreen needleleaf trees) | |
| ▤ Mediterranean scrub and woodland | |
| ▤ temperate grassland | |
| ⣿ desert (+some semi–arid communities) | |
| ▥ tropical savanna | |
| ▦ tropical woodland and thornscrub | |
| ▥ tropical rain forest | |
| ▥ tropical deciduous forest | |

**Fig. 8.4 Major terrestrial biomes of the world**

## THE TROPICAL RAIN FOREST, EQUATORIAL FOREST BIOME

### Characteristics

a) Nearly all evergreen;
b) Leathery leaves;
c) Buttress roots;
d) Cauliflowery flowers on trunk;
e) **Epiphytes** – plants on trunks;
f) **Lianas** – climbers;
g) **Saprophytes** – plants in well-shaded areas growing without photosynthesis;
h) Little ground-level vegetation;
i) Variety of species;
j) Large **biomass** – very high net primary productivity, much in the tree layers;
k) Large litter supply which decays rapidly and is destroyed more rapidly than supplied;
l) Soils poor because nutrients circulated in vegetation.

### Adaptation

a) Traditional **slash and burn** farming releases nutrients to soil – some fertility – soon lost therefore people move on – area returns to **secondary forest**;

b) If population pressure reduces interval before people return to the original sites of slash and burn cultivation, then nutrient levels decline and ability to maintain population declines;

c) Commercial farming and ranching – some adaptation merely involves **simplification**, for example rubber plantations – destroys species variety. Others create a new agricultural ecosystem, for example cattle ranching.

### Consequences

a) **Soil erosion** because less interception and no roots to bind soil – river sediment loads and floods increase;

b) **Greenhouse effect** accelerates because removal of biomass increases $CO_2$, photosynthesis is reduced, therefore $CO_2$ builds up (see chapter 7);

c) Areas return to **secondary forest** which is lower, contains smaller trees, has more undergrowth and will be poorer in species. The new trees grow quickly but the wood has few uses compared to hardwoods of primary climax forests.

The tropics are a zone where unwise human actions can have devastating consequences.

## BOREAL CONIFEROUS FOREST BIOME

66 Characteristics of the boreal coniferous forest biome. 99

### Characteristics

a) **Evergreen** conifers;

b) **Needle** leaves;

c) **Shaped** to shed snow/weight;

d) Continuous **stand**;

e) No lower layers;

f) Uncontinuous **ground layer** of lichens, mosses and bog moss dependent on light/drainage conditions;

g) Relatively few species;

h) Primary productivity: 35 per cent of rain forest and 66 per cent of deciduous forest because low energy inputs;

i) **Slow decay** of litter therefore much of nutrient stored here or in biomass;

j) Species distribution controlled by: climate, soil and people;

k) **Podsol soils** (chapter 9).

### Adaptation

a) Burning – lightning strikes in summer – litter layer ideal for spreading fire. Careless human actions may also cause fire. Slow recovery time because low temperatures in winter.

b) Clearance for mining, settlement, transport and in some cases – for example parts of Peace River country, North Alberta – agriculture.

c) Commercial timber production for timber and paper pulp – very controlled with replanting programmes common.

### Consequences

a) Environmental damage from exploitation has been slowed and now carefully managed exploitation ensures replanting;

b) **Acid rain** (chapter 7) – not a product of exploitation, but does have bad effects killing new shoots, slowing growth rates and killing needles. Seventeen per cent of Swedish forest is affected (worse effects are found in West German forests although these are not always in the boreal biome).

## CASE STUDY: MANAGING A THREATENED FOREST ECOSYSTEM – THE ALGONQUIN

The Forestry Authority was established by the Ontario government in Canada in 1975 to manage the exploitation of the resources of the 2 million-acre Algonquin Provincial Park. The park was established as a game preserve in 1893 because of the effects of hunting on the game and bears. In this way it was hoped that dynamic equilibrium could be maintained between the need to supply logs to timber and paper mills dependent on the area, the yield capacity of the area, and the recreational demands of visitors. People were aware that without management of the resources the 130-year-old timber industry might destroy the ecosystem of the boreal forest.

The Authority prepares forest-operating plans for the extraction of 17 million cubic feet of timber felling and the replanting of these areas. Timber extraction is banned in 25 per cent of the park which is zoned as primitive, natural, historic and recreational areas. Timber extraction is placed alongside recreation as an equally important use.

Recreational use is also zoned, with the most intensive uses taking place along the one 50-km public road which traverses a corner of the park. Here are eight trails to introduce visitors to the breadth of the forest's ecology (see the case study of the Spruce Bog succession, above). Beyond the trails, areas are zoned for different outdoor pursuits from white water canoeing to trekking in remote areas.

The Golden Lake Indians are the only people permitted to trap in the park, and then only in restricted areas. In this way timber production directly employs 3200 workers and 10 000 more in associated activities in Ontario which depend on the resource, and yet more work for the Ministry of Natural Resources who administer the recreational activities.

## TEMPERATE DECIDUOUS FOREST BIOME

> **Characteristics of the temperate deciduous forest biome.**

### Characteristics

a) **Deciduous** with some needle leaves – people have introduced coniferous into many areas for commercial reasons;
b) **Broad** leaves;
c) Climbers present;
d) **Epiphytes** on trunks;
e) **Species density** low'
f) Primary productivity about half of rain forest;
g) **Shrub layer** can develop due to light penetration;
h) Brown forest soils develop;
i) Nutrients stored in soil and biomass;
j) Breakdown of litter efficient.

### Adaptation

a) Cleared over the centuries for agriculture – remnants in hedgerows and protected forests (Forest of Bere, Hampshire);
b) **Coppiced** for fuel in past;
c) Frequently replanted with faster-growing coniferous woodland for commercial reasons.

### Consequences

a) **Conservation** of deciduous woodland and **replanting** for aesthetic reasons;
b) Loss of species – can be caused by diseases, for example, Dutch Elm disease which destroyed much of one species in the 1970s;
c) Grassland can revert to woodland if natural controls removed, for example rabbits killed in 1954 myxamatosis outbreak – no animal to eat new saplings and shrubs so **reversion** to woodland on, for example, South Downs escarpment.

## TROPICAL SAVANNA BIOME

> **Characteristics of the tropical savanna biome.**

### Characteristics

a) Tall **grassland** with variable amounts of evergreen trees, for example eucalyptus in Australia;
b) Marked seasonality of climate – hence adaptation of species – **xeromorphic** or drought-tolerant;
c) Primary productivity at same level as deciduous forests, i.e. half of rainforest;
d) Partial/total loss of loess;
e) Large herbivore mammals;
f) **Fire**, both natural and people-made, helps maintain the biome – much vegetation is fire resistant, for example Black Boy bushes in Australia;
g) Associated with plains and plateaux;
h) Only two layers;
i) Probably *not* a climax vegetation.

### Adaptation

a) **Fire**, as noted above – promotes grass growth rather than trees so favoured by herdsmen – increases nutrient levels;

b) Thorny plants replace woodland after fire;

c) **Human intervention** due to population pressure, particularly on desert margins – excessive grazing by cattle and goats – increased gathering of timber for cooking fuel may lead to **desertification** (chapter 7);

d) Commercial agriculture may also **simplify** the biome.

### Consequences

a) Where commercial farming attempted, loss of soil capability very rapid, for example groundnut scheme in Kenya, 1947;

b) Desertification (see chapter 7);

c) Conservation of fauna and flora in game reserves.

## TEMPERATE GRASSLAND BIOME: PRAIRIES, PAMPAS, VELD OR STEPPE GRASSLAND

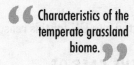
**Characteristics of the temperate grassland biome.**

### Characteristics

a) **Grassland** with some trees – aspen in North America.

b) Probable role of natural and people-made fire in denuding;

c) Probable role of large herbivores – **bison** (North America) in maintaining grassland;

d) Nutrients stored overwhelmingly in soil; Low primary productivity (half deciduous forest and two-thirds of savanna);

e) High humus, **black earth** or **chernozem soils**;

f) Aspen grown in an **ecological niche** – area best suited to an organism, i.e. dusty basins caused by wallowing bison were beds for seed deposited by birds.

### Adaptation

a) People probably used fire to clear areas – lightning fires did the same;

b) Hunting of bison in eighteenth/nineteenth centuries for fur, hide and meat – almost extinct by twentieth century – **dominant herbivore** lost;

c) Cattle ranching replaced dominant herbivore;

d) Ploughing for cereal cultivation – elimation of insects and seeds – **simplification** – nutrient cycle broken.

### Consequences

a) Some areas under cultivation were more susceptible to wind and water erosion (after ploughing) – 'Dust Bowl' in 1930s the best case;

b) To retain fertility, measures needed – rotation, fertilisers;

c) Belated attempts to conserve the large herbivores;

d) Loss of variety of insect life.

**NB** There are other biomes which are included in some syllabuses: Mediterranean, Desert, Polar (tundra), Mountain. You might need to have prompts for these under the same headings as used here.

## HALOSERES

A halosere develops in saline conditions.

## MANGROVE OR TIDALLY SUBMERGED WOODLAND

**Aspects of the mangrove halosere.**

### Characteristics

a) Salt-tolerant trees;

b) Up to 30° N and S;

c) Low wave action, therefore muds accumulate;

d) Stilt roots hold sediments;

e) May extend up estuaries;

f) Provides an ecological niche for breeding fish.

### Issues

a) Clearance for tourist activities;

b) Effects of marine pollution;

c) Loss of niche;

d) Need to conserve.

## CORAL

Aspects of the coral halosere.

### Characteristics

a) Can be **fringing** or **barrier** or **atoll**;

b) Tropical;

c) Sea temperature 21°C and above;

d) Platform for growth;

e) Little sediment to cloud water;

f) Common on east coasts of continents;

g) Dominated by carnivorous animals feeding on **zooplankton**;

h) Animals build an external skeleton to give solidity to reefs.

### Issues

a) Destruction for navigation;

b) Destruction by pollution;

c) Removal for souvenirs, etc. – black coral may not be imported into UK for this reason.

## CASE STUDY: TIDAL MARSHLAND HALOSERE – THE CASE OF LANGSTONE HARBOUR, HAMPSHIRE

### Characteristics

a) In area sheltered from wave action – estuary, lee of a spit or barrier where **accretion** takes place;

b) Vegetation promotes **accretion**;

c) Vegetation adapted to salt/**inundation**;

d) New areas have little vegetation cover;

e) Older areas are densely covered;

f) Tidal inundation declines as marsh level rises;

g) Marshes have a **creek system** and **pans** or small basins;

h) There is a **succession** of vegetation as accretion occurs – all plants must be tolerant of a wide range of conditions;

i) Estuaries and deltas very **productive ecosystems** – nutrient-rich;

j) Much modified by people – 75 per cent estuary wetlands altered.

### The succession

1 **Algae**, for example, enteromorpha;

2 Salt-tolerant **halophytes**, for example salicornia – annuals which manage to trap some silt;

3 Less salt-tolerant, for example, *Aster tripolium* and *Puccinellia maritima* with **fleshy foliage**;

4 **Creeks** have own **niche** on banks, for example, *Halimone portulacoides* – may aid **levée** development;

5 Marsh grass, *Juncus maritimus*;

6 When little inundation, reeds and sedges;

7 In UK, introduction since 1870s of spartina has speeded up accretion.

### Uses of Langstone Harbour

a) The **conservation** of marshland as wintering area for, for example, Brent Geese;

b) The **reclamation** of marine inlets by waste dumping – reclaimed former saltings, i.e. salt-drying pans;

c) The presence of major built-up areas, playing fields, roads along harbour edges leading to:

d) Sewage disposal – untreated offshore beyond harbour mouth and in harbour at Budds Farm;

e) Recreational use, fishing, sailing, water-skiing – marinas;

f) Industrial use – offloading quays for dredged offshore gravel;

g) The maintenance of a nature reserve at north of harbour;

h) Dune landscapes on East Winner.

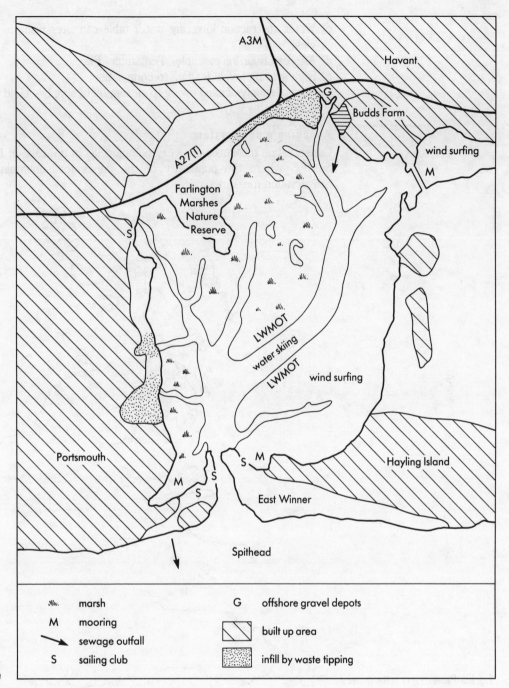

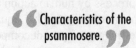
Fig. 8.5 Langstone Harbour, Hampshire

---

## PSAMMOSERE – THE SAND-DUNE ECOSYSTEM

" Characteristics of the psammosere. "

### Characteristics

a) Areas with large tidal range;

b) Low gradient beach;

c) Suitable low-lying area behind beach;

d) **Frontal** dune grass around litter or vegetation – couch grass, sea rocket, lyme grass;

e) Grass root system holds sand;

f) **Marram grass** main stabilising agent with dune fescue, bindweed and seaholly also tolerating water deficiency and nutrient deficiency of dunes;

g) Calcareous dunes have richer vegetation than silica sand and older leached dunes. Latter colonised by acid-loving heathland plants – heathers and ling;

h) Grazing by rabbits can reduce plant varieties;

i) Slacks develop own niche and fauna;

j) Further inland pine may seed or be planted;

k) Fauna depends on the insect and small mammal life of dunes.

### Adaptation

a) By erosion – marine transgression probably aided by:

b) Human action re-exposing sand, for example, barbeques, horse-riding, dune buggies;

c) Water abstraction lowering water table can alter niche of slacks and reduce the food web;

d) Afforestation, for example, Tentsmuir, Fife;

e) High grazing levels leading to compaction;

f) Loss of supply of sand owing to cutting off of supply – Gold Coast, Queensland, has lost dunes in this way.

### Managing a dune system

Conservation measures are very common, as in the case of East Head, West Wittering, Sussex, where the popularity of the dunes for sunbathing has necessitated strong conservation measures.

Fig. 8.6 Psammoseres in the United Kingdom

## EUTROPHICATION

Eutrophication is the natural process of ageing of an aquatic ecosystem through enrichment by nutrients. This occurs in lakes because they are fed by nutrient-rich streams and, as lakes age, plant debris provides further nutrients. **Oligotrophic** lakes are the other extreme, where nutrient cycling and productivity are low, for example high altitude lakes. **Mesotrophic** refers to the intermediate stage between these extremes of nutrient cycling. **Cultural eutrophication** is the acceleration of the process by human action and is the most common form of **aquatic pollution**. Mesotrophic and oligotrophic aquatic systems can 'cleanse themselves' because oxygen levels have the capacity to decompose waste. In industrial societies (mainly) industrial, agricultural and city wastes, high in organic substances, are discharged into rivers. Decomposition takes more oxygen and depletes supply of oxygen for other aquatic life. **Biochemical oxygen demand** (BOD) is a measure of the oxygen needed to decompose wastes in water. Higher nutrient levels (phosphates and nitrates) cause plants/algae to increase, and as these die they starve the system of oxygen.

### Sources

a) Domestic sewage, particularly from detergents;

b) Urban run-off which includes animal waste, garden fertilisers and oil products;

c) Industrial wastes, for example, paper mills;

d) Nitrates in run-offs from less well-managed farmland – most found with irrigation;

e) Slurry – i.e. animal waste from intensive farming draining directly into a stream.

### Environmental effects

a) Variety of organisms reduced and can result in 'death' of a lake – only fungi and bacteria remain – 25 per cent of lakes in Sweden are dead;

b) Hindered shoreline activity and even declining property values;

c) Smell and taste to water, reducing recreational value and water consumption potential.

### Governmental concern and action

Rules on sewage emissions in West Germany – concern over killer algae in Sweden. Green Party votes climb – 1988, 5.8 per cent in Sweden, 8 per cent in West Germany.

## POLLUTION

## ACCUMULATION

Pollution by toxic substances can pass through trophic levels, and if not broken down or excreted they concentrate in the higher levels of a food chain. Accumulation is worse in aquatic chains where there are more trophic levels to enable them to concentrate. Pesticides used in agriculture, for example DDT (dichloro-diphenyl-trichloro-ethane), have had this effect and have been banned in many countries (Fig. 8.7).

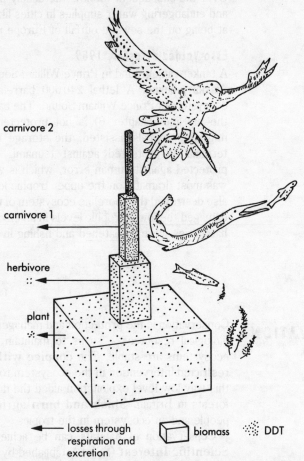

carnivore 2

carnivore 1

herbivore

plant

losses through respiration and excretion

biomass

DDT

Fig. 8.7 The accumulation of DDT in a foodchain

## CASE STUDY: MINAMATA – THE TRAGEDY OF FOOD CHAIN ACCUMULATION

Minamata Bay was a source of fish and shellfish for the Japanese living in the surrounding villages. Also, on the shore of the bay the Chisso Corporation had located a chemical plant. In the 1950s, mammals showed signs of unusual behaviour, and then people began to have headaches, blurred vision and speech impediments. Over 45 people died after violent fits. The cause was found to be discharges of mercury chloride by Chisso, which were converted in the sea to methyl mercury, a highly toxic compound which can enter the food chain. The government recognised 112 cases in 1968. The company did not accept responsibility until 1973, when forced by government action. In all, 35 000 people suffered Minamata disease – a disease of accumulation in a food chain.

## THERMAL POLLUTION

Thermal pollution originates mainly from power stations and from individual plants. Water is used to cool steam-driven turbines before it is returned to a river, lake or the sea. Warmer water increases dissolved oxygen demand and kills aquatic species. A quarter of US run-off is affected and this could increase.

## SPILLS AND ECOSYSTEMS

### Amoco-Cadiz disaster, 1978

200 000 tonnes of oil were released from this wrecked tanker onto 200 km of Brittany coastline. The oil slick extended 60 km out to sea. The cost to clear it up totalled $75 million. Oil destroys natural insulation and buoyancy in birds and mammals, kills shellfish and so destroys the livelihood of fishing villages.

### Sandoz disaster, 1986

Following a fire, toxic chemicals were accidentally washed into the river Rhine at Basle. Over the subsequent months the deadly flow travelled down the river, killing aquatic life and endangering water supplies in cities like Cologne and on the delta, and Dutch concern at being on the sewage outfall of Europe rose.

### Esso Valdez disaster, 1989

A tanker ran aground in Prince William Sound, just after leaving the Valdez terminal of the Alaska pipeline. A lethal 240 000 barrels of crude oil were released into the fragile ecosystem of Prince William Sound. The bay's ecosystem, unlike those along the course of the pipeline (chapter 6) is not protected. Permafrost dangers have been overcome, migrating animals assisted, the storage tanks can cope with 4.25 m of snow and the terminal is protected against tsunami. Unfortunately, the ocean approach was not protected against human error, which is what caused the disaster. The ecological effect was most dramatic on the upper trophic levels, seals, whales, sea otter, but the oil slick also destroyed the shoreline ecosystem of the islands and, together with the dispersants, it damaged the lower trophic levels in an environment which will be slow to recover. Salmon hatcheries were threatened and fishing livelihoods destroyed.

## CONSERVATION

Conservation is the protection and management of a resource – in this context, renewable biological resources – so as to maintain its utility. This is achieved by ensuring that ecosystems are stable. The **coppice with standard** system of woodland management restricted intervention in the ecosystem to a level that enabled recovery to take place. In this way low-level exploitation aided the development of a diverse food web in deciduous forests in Britain. **Slash and burn agriculture** also enabled a similar coexistence of people and an ecosystem in the tropics.

Conservation of habitats can be achieved by law, for example **Sites of Special Scientific Interest** (SSIs) established by the National Parks and Access to Countryside Act 1949 and the Wildlife and Countryside Act 1981. Flora and fauna (also geological and physiographic features) are conserved by agreement between land-owners, occupiers and the county council. There are over 4000 SSIs covering 13 669 sq km and their protection is by agreement. The Havergate marshes on the Norfolk Broads were saved from reclamation for arable agriculture and kept as drained pasture by grants to enable the marshes to be kept in their former use: farmers were paid to maintain the old ecosystem.

In Austria farmers are paid to maintain the traditional alpine landscape (ecosystem) rather than abandon it. The reasons here are more to do with tourism than conservation per se.

Conservation of wilderness areas in the UK led to the **National Parks**, most of which are people-made landscapes. **National Nature Reserves** (182 of 1400 sq km, for example Crymlyn Bay, Swansea, and Farlington Marshes, Portsmouth) and **Local Nature Reserves** are further examples of conservation strategies.

## VEGETATION MAPPING

**Aspects to measure**

a) **Stratification;**

b) **Horizontal distribution;**

c) **Abundance of species.**

Stratification involves drawing plans of 100–500 sq m area, measuring height, width and height to branches and lower crowns of trees, and obtaining mean measurements of all of these. The lowest layers can be examined by **quadrat sampling**. Quadrats can be used systematically on transects (up a slope, for instance), or randomly. Quadrat size depends on size of vegetation units. Watch for **edge effect** if frame too small. Vegetation is recorded in each quadrat so that frequencies are built up for different environments, for example north- and south-facing slopes on a particular rock type. **Frequency** recorded is either **local shoot**, i.e. measures cover, or **local rooted**, i.e. measures density, or both. Cover and occurrence can use **Braun-Blanquet rating system**.

It enables one to determine **density** – the number of species per unit area and **cover** – the percentage of the ground overlain by crowns or canopies of plants.

The mapping would enable you to relate the vegetation patterns to aspect, relief, slope, geology, soil and time since human interference.

# PROJECTS

- The study of plant associations in deciduous and coppiced woodland.
- Vegetation and aspect, slope, soils and geology.
- The effects of sewage outfalls, industrial plant and power stations on the life in a river.
- The impact of walkers, riders on the vegetation along a footpath.
- The patterns of vegetation on dunes/marshland and people's effects on the system and its management.
- Vegetation succession on a spoil tip or rubbish dump – natural or managed succession.

# EXAMINATION QUESTIONS

### Question 1

a) Describe and compare the nutrient cycle of a tropical evergreen forest with that of temperate grasslands.                                                                      *(10)*

b) Discuss how differences in the character of the nutrient cycles between these two biomes help to explain differences in soil type and soil fertility between the regions occupied by the two biomes and in their agricultural potential.                    *(10)*

c) Explain why concern about the destruction of tropical forests extends to countries outside the tropics.                                                                           *(5)*

(Oxford, Specimen Question, 1990)

### Question 2

Discuss the criteria and methods you would use to make a comparative field study of TWO ecosystems.

(WJEC, 1988)

### Question 3

Study the information given in Figs 8.8 and 8.9 which refer to the southern Lake Michigan area of the USA.

Fig. 8.9 shows the vegetation and soils found on the sand dunes on the southern shores of the lake, together with a table showing the difference in the uptake of plant nutrients by deciduous woodland and coniferous forest.

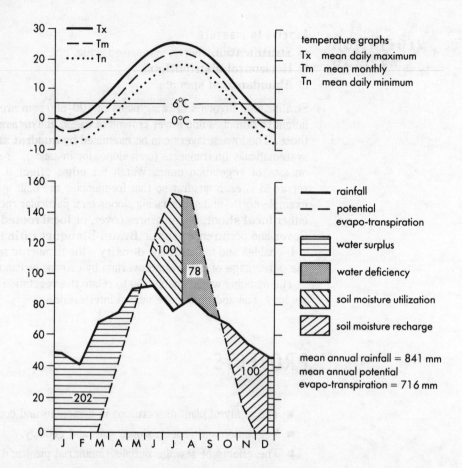

Fig. 8.8 Climate graphs

temperature graphs
Tx  mean daily maximum
Tm  mean monthly
Tn  mean daily minimum

rainfall
potential evapo-transpiration
water surplus
water deficiency
soil moisture utilization
soil moisture recharge

mean annual rainfall = 841 mm
mean annual potential
evapo-transpiration = 716 mm

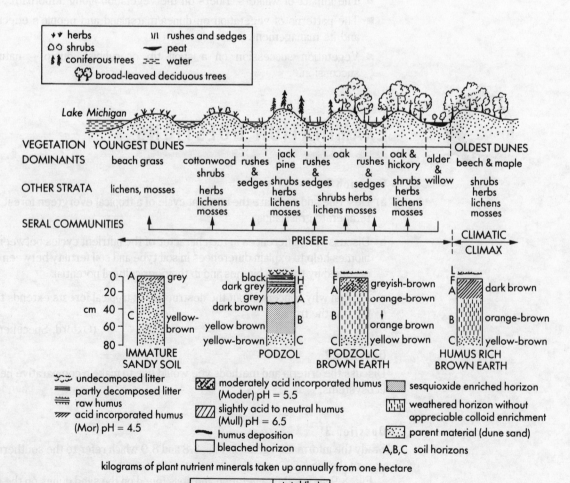

Fig. 8.9 Plant succession and soil development on sand dunes

herbs
shrubs
coniferous trees
broad-leaved deciduous trees
rushes and sedges
peat
water

VEGETATION
DOMINANTS

OTHER STRATA

SERAL COMMUNITIES

YOUNGEST DUNES

beach grass

lichens, mosses

cottonwood
shrubs

herbs
lichens
mosses

rushes
&
sedges

jack
pine

shrubs
herbs
lichens
mosses

rushes
&
sedges

oak

shrubs
herbs
lichens mosses

rushes
&
sedges

oak &
hickory

shrubs
herbs
lichens
mosses

alder
&
willow

beech & maple

shrubs
herbs
lichens
mosses

OLDEST DUNES

PRISERE

CLIMATIC
CLIMAX

IMMATURE
SANDY SOIL

grey

yellow-brown

PODZOL

black
dark grey
grey
dark brown
yellow brown
yellow-brown

PODZOLIC
BROWN EARTH

greyish-brown
orange-brown
orange brown
yellow brown

HUMUS RICH
BROWN EARTH

dark brown
orange-brown
yellow-brown

undecomposed litter
partly decomposed litter
raw humus
acid incorporated humus (Mor) pH = 4.5

moderately acid incorporated humus (Moder) pH = 5.5
slightly acid to neutral humus (Mull) pH = 6.5
humus deposition
bleached horizon

sesquioxide enriched horizon
weathered horizon without appreciable colloid enrichment
parent material (dune sand)

A,B,C soil horizons

kilograms of plant nutrient minerals taken up annually from one hectare

|  | total (kg) |
| --- | --- |
| deciduous forest | 430 |
| coniferous forest | 225 |

Fig. 8.8 is a climate graph for a weather station in the area.

a) Define the following terms as used in Fig. 8.9:
   i) seral community;
   ii) prisere;
   iii) climatic climax community.                                                     (6)

b) Use the information given on the graph (Fig. 8.8) to explain why the climatic climax
   vegetaton of the southern Lake Michigan area is deciduous rather than boreal
   coniferous forest or temperate grassland.                                           (6)

c) What evidence is there on Fig. 8.9 that each plant community, in time, creates
   conditions which favour more complex and demanding communities?                     (6)

d) Assume that some of the oldest dunes at present covered by beech and maple woodland
   with a brown earth soil were to be cleared and replaced by permanent pasture for
   sheep.
   i) What THREE changes might occur over a period of many years in the soil profile?  (3)
   ii) Explain why you would expect any TWO of such changes to occur.                  (4)

(COSSEC, Sample Paper, AS, 1990)

## Question 4

This is a decision-making exercise.

Study the sketch-map, Fig. 8.10, showing Oxwich National Nature Reserve (ONNR) in
South Wales.

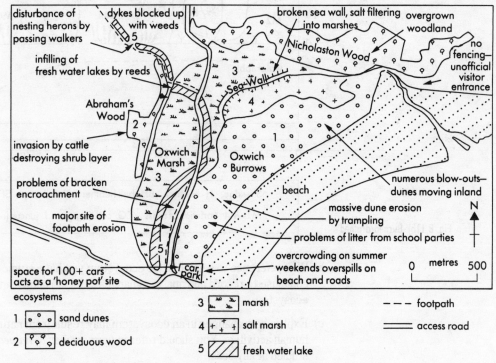

Fig. 8.10    Source: Ecology and Conservation Unit, University College, London

a) Summarise a field-study programme that would enable you to investigate the structure
   of ONE of the small-scale ecosystems shown on the map.                             (6)

b) With the aid of a labelled diagram, or diagrams, show how human activities can affect
   the structure and functioning of any small-scale ecosystem you have studied.       (8)

(London, AS 16–19 Sample)

## Question 5    Ecosystems and human activity

Study Figs 8.11 a and b which show the effects of burning and draining on a moorland
ecosystem in northern Britain.

a) Describe, and suggest reasons for, the changes in the structure and composition
   of the moorland ecosystem after burning and draining.                             (7)

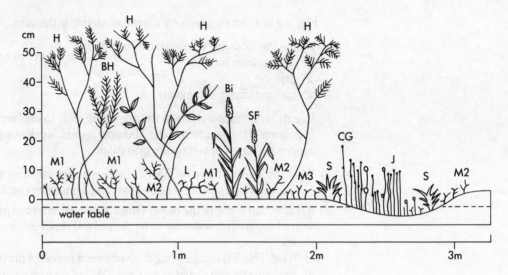

Fig 8.11a) Unburned/undrained

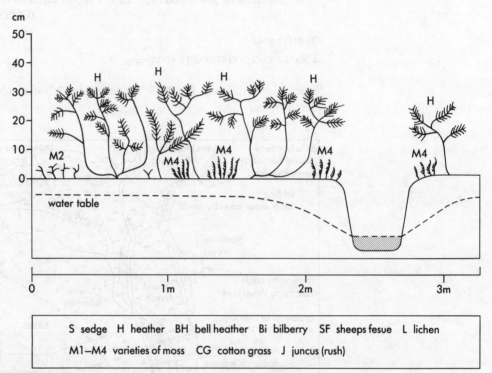

| S sedge | H heather | BH bell heather | Bi bilberry | SF sheeps fesue | L lichen |
|---------|-----------|-----------------|-------------|------------------|----------|
| M1—M4 varieties of moss | | CG cotton grass | | J juncus (rush) | |

Fig. 8.11b) Burned/drained

b) With reference to a small-scale ecosystem you have studied, outline the methods you could use to record and analyse changes in the structure and composition of the ecosystem. *(9)*

c) Explain how changes in an ecosystem may result from natural processes and/or human activities. You should refer in your answer to *named* examples you have studied at any scale. *(9)*

(London 16–19, 1989)

## Question 6

This is a data response question.

Fig. 8.12 is a model of the mineral nutrient cycle. The model is illustrated in three different environments shown in Fig. 8.13.

a) Use your knowledge of the three environments to explain why:
  i) plant litter is the smallest store of mineral nutrients in the temperate deciduous forest environment (Fig. 8.13 a);
  ii) the soil is the largest store of mineral nutrients in the hot desert environment (Fig. 8.13 b);
  iii) the transfers between stores are so similar in the savanna environment (Fig. 8.13 c). *(15)*

b) Using Figs 8.14 and 8.15 which show the effects of fire on a savanna environment and Fig. 8.13 c, describe and explain the differences between a savanna region subject to regular controlled burning and a savanna region subject to a major natural fire in:
  i)  the composition of the vegetation,
  ii) the mineral nutrient cycle.

(10)
(JMB, 1988)

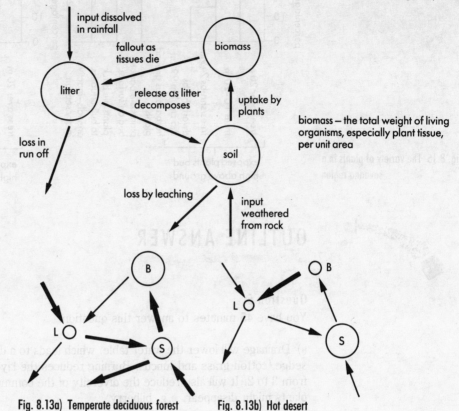

input dissolved in rainfall

fallout as tissues die

biomass

litter

release as litter decomposes

uptake by plants

loss in run off

soil

biomass – the total weight of living organisms, especially plant tissue, per unit area

loss by leaching

input weathered from rock

Fig. 8.12 A model of the mineral nutrient cycle

B

L

S

Fig. 8.13a) Temperate deciduous forest

B

L

S

Fig. 8.13b) Hot desert

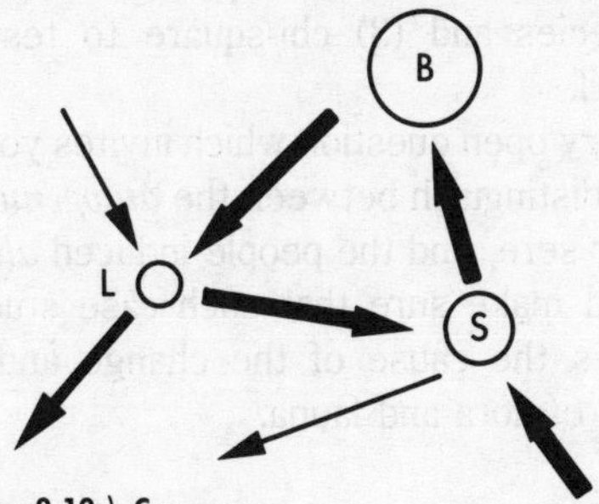

B

L

S

B  biomass
L  litter
S  soil

The size of the nutrient store symbol (B, L, S) is proportional to the quantity of nutrients stored. The thickness of the transfer arrows is proportional to the amount of nutrients transferred.

Fig. 8.13 Mineral nutrient cycles    Fig. 8.13c) Savanna

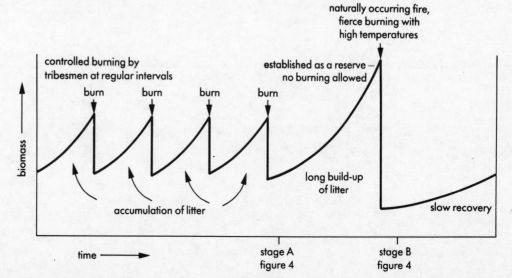

naturally occurring fire, fierce burning with high temperatures

controlled burning by tribesmen at regular intervals

burn    burn    burn    burn

established as a reserve — no burning allowed

biomass

accumulation of litter

long build-up of litter

slow recovery

time

stage A figure 4

stage B figure 4

Fig. 8.14 The effect of burning on an area of savanna vegetation

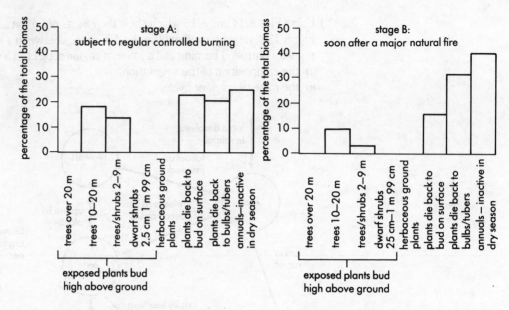

Fig. 8.15 The variety of plants in a savanna region

## OUTLINE ANSWER

### Question 5
You have 45 minutes to answer this question.

a) Drainage will lower the water table, which leads to a decline in marshy species such as sedge, cotton grass and Juncus. Burning reduces the layers or stratification of the plants from 3 to 2. It will also reduce the diversity of the communities at ground level and other plants might disappear, e.g. bilberry.

b) Here the opportunity is yours to discuss a small-scale ecosystem and its changes. Why not use the saltmarsh or sand-dune changes, but keep to the small scale. Remember to outline the methods you would use: (1) quadrat sampling to obtain the data, (2) diagrams to show dominant species and (3) chi-square to test the association between species, structure, slope, soil.

c) This is again a very open question which invites you to perhaps use your own fieldwork. The answer should distinguish between the *autogenic* changes of a succession leading to a climax vegetation or sere, and the people-induced *allogenic* changes, such as flooding and burning. You should make sure that each case study (perhaps two or three would be appropriate) includes the cause of the change and the impact of that change on the distribution and mix of flora and fauna.

## CHAPTER 9

# SOILS

## GETTING STARTED

Soils are an integral part of the biosphere and should be studied in conjunction with ecosystems. Several syllabuses are not very specific about the content of this part of your studies, but past papers might help you to fill the gaps. In at least two cases the regions which you select will determine which soils are studied (and which ecosystems). If you are doing the Oxford and Cambridge syllabus then soils might be beyond your studies.

**Soil** is composed of:
a) Inorganic material, rock particles; individual minerals.
b) Organic material in partially decomposed material.
c) Soil, water and air plus soil organisms.
All of these are held together as **aggregate units** or **peds** or **structural units**. It is a three-dimensional body, organised vertically as a **soil profile** with **soil horizons** and horizontally into areas of similar horizon characteristics.

**BASIC SOIL PROCESSES**

**SOIL CLASSIFICATIONS**

**TRANSFORMATIONS**

**PROCESSES OF PEDOGENESIS**

**ANALYSING SOILS**

**AGRICULTURE AND SOIL MODIFICATION**

# ESSENTIAL PRINCIPLES

**BASIC SOIL PROCESSES**

Four basic **processes** operate on soils (Fig. 9.1). Water is essential for the soil-forming processes. **Pedogenesis** depends on water held in the **pores** or moving through them, in the processes outlined below.

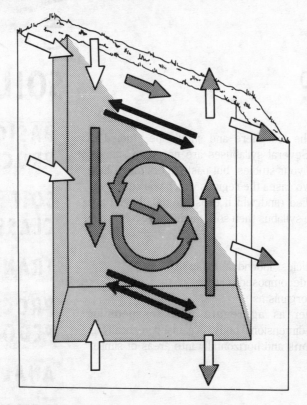

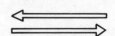
Transformation processes
Internal reorganisation of matter and redistribution of energy, but *in situ* e.g. decay of organic matter, weathering of primary and secondary minerals. Net loss of mass and free energy accompanies such processes

Transfer processes
Internal reorganisation of matter and redistribution of energy, but involving movement e.g. translocation of iron, clay, humus, and hydrated ions, diffusion of gases, ion exchange, mass-movement and through flow, capillary rise, mixing by soil fauna, cryoturbation

Input processes
Inputs of mass, (e.g. organic matter as litter, rainwater, respiratory $CO_2$, regolith by weathering, mass movement and through flow from upslope) and energy as chemical, kinetic, radiant, or mechanical energy or as some combination

Output processes
Many similar to input processes e.g. downslope mass movement and through flow, deep percolation and leaching. Some uniquely output processes such as water and nutrient uptake by plant roots

**Fig. 9.1 Four basic soil processes**

## Inputs

a) Water and dissolved substances released from weathering and decomposition;
b) Organic matter;
c) Minerals from erosion and streams flooding;
d) Energy.

## Outputs

a) Materials removed in solution or washed out;
b) Material removed by erosion;
c) Decomposition losses.

## Transfers or translocations

These constitute the reorganisation of energy. Horizons that lose material are **eluvial** whereas those which gain matter are **illuvial**. Only on a slope is a transfer horizontal.

## Transformations

Transformations involve **decomposition** to form **humus** and **weathering** to form the secondary minerals – clay.

## PEDOGENESIS

**Pedogenesis** depends on the following factors (the Jenny model):

1 inorganic;                    4 relief or topography;
2 organic;                      5 time;
3 climate;                      6 people.

The interaction of these factors in different locations results in the soils characteristic of different ecosystems and biomes.

## SOIL CLASSIFICATIONS

*Different soil taxonomies.*

Soils may be classified, and you will find two systems in use:
- The FAO World Soil Taxonomy (Table 9.1)
- The US Soil Taxonomy (Table 9.2)

Sometimes a third system is mentioned, the British System (Table 9.3) though this is essentially a variant of the US Soil Taxonomy.

**NB. Do not worry about all the official terms here.**

| | | |
|---|---|---|
| 1 | Fluvisols | Soils in alluvium (or colluvium) (Fluvents) |
| 2 | Gleysols | Hydromorphic soils in non-alluvial materials |
| 3 | Regosols | Weakly developed soils in loamy or clayey unconsolidated deposits other than alluvium |
| 4 | Lithosols | Shallow soils with hard rock at less than 10 cm depth |
| 5 | Arenosols | Soils with unconsolidated sandy deposits other than alluvium |
| 6 | Rendzinas | Soils with a mollic A horizon over extremely calcareous material (Rendolls) |
| 7 | Rankers | Soils with an umbric A horizon and no subsurface diagnostic horizon |
| 8 | Andosols | Soils, mainly in volcanic ash, with a low bulk density and an exchange complex dominated by amorphous material (Andepts) |
| 9 | Vertisols | Cracking clay soils (Vertisols) |
| 10 | Solonchaks | Saline soils in non-alluvial materials |
| 11 | Solonetz | Soils with a natric B horizon |
| 12 | Yermosols | Desert soils with a very weak ochric A horizon (containing little organic matter) (Typic Aridosols) |
| 13 | Xerosols | Desert soils with a weak ochric A horizon (containing more organic matter than Yermosols) (Mollic Aridosols) |
| 14 | Kastanozems | Soils with a mollic A horizon and a calcic or gypsic horizon or concentrations of lime (Ustolls, or Chestnut soils of Thorp and Smith 1949) |
| 15 | Chernozems | Soils with a mollic A horizon and a calcic or gypsic horizon or concentration of lime |
| 16 | Phaeozems | Soils with a mollic A horizon and no calcic or gypsic horizon or concentrations of lime |
| 17 | Greyzems | Soils with a mollic A horizon and bleached coastings on ped faces of subsurface horizons: usually with an argillic B horizon |
| 18 | Cambisols | Soils with a cambic B horizon and without horizons diagnostic of other classes |
| 19 | Luvisols | Soils with an argillic B horizon of medium to high base status |
| 20 | Podzoluvisols | Soils having an argillic B horizon with an irregular upper boundary due to a tonguing E horizon or to ferruginous nodules |
| 21 | Podsols | Soils with a spodic B horizon (Spodosols) |
| 22 | Planosols | Soils with a gleyed albic E horizon over a slowly permeable horizon |
| 23 | Acrisols | Soils with an argillic B horizon of low base status |
| 24 | Nitosols | Soils with an argillic B horizon extending to 150 cm or more |
| 25 | Ferralsols | Soils with an oxic B horizon (Oxisols) |
| 26 | Histosols | Peat soils (Histosols) |

Table 9.1 FAO World Soil Taxonomy

1 **Entisols** Weakly developed mineral soils
  1.1 Aquents – with gleyic (hydromorphic) features
  1.2 Arents – others with fragments of soil horizons (disturbed soils)
  1.3 Psamments – others in sandy materials
  1.4 Fluvents – others in alluvium or colluvium
  1.5 Orthents – other Entisols
2 **Vertisols** Cracking clay soils
  2.1 Xererts – with long dry periods
  2.2 Torrerts – usually dry
  2.3 Uderts – usually moist
  2.4 Usterts – others, with short dry periods
3 **Inceptisols** Moderately developed soils of humid regions normally with a cambic horizon and no spodic, argillic or oxic horizon
  3.1 Aquepts – with gleyic (hydromorphic) features
  3.2 Andepts – others mainly in volcanic ash with a low bulk density and an exchange complex dominated by amorphous material
  3.3 Plaggepts – others with a plaggen epipedon
  3.4 Tropepts – others of tropical climates, mainly with ochric and cambic horizons
  3.5 Ochrepts – others of mid to high latitudes, mainly with ochric and cambic horizons
  3.6 Umbrepts – others of humid mid to high latitudes, mainly with an umbric epipedon
4 **Aridisols** Soils of deserts and semi-deserts
  4.1 Argids – with an argillic or natric horizon
  4.2 Orthids – others, without an argillic or natric horizon
5 **Mollisols** Base-rich soils with a mollic epipedon (mainly of the steppes)
  5.1 Abolls – with albic and argillic horizons and affected by fluctuating ground-water
  5.2 Aquolls – others with gleyic (hydromorphic) features
  5.3 Rendolls – others with a mollic epipedon over extremely calcareous materials
  5.4 Xerolls – others with long dry periods
  5.5 Borolls – others of cold climates (with a frigid, cryic or pergelic temperature regime)
  5.6 Ustolls – others of sub-humid or semi-arid climates

  5.7 Udolls – others of humid climates
6 **Spodosols** Soils with a spodic horizon (most podsols)
  6.1 Aquods – with gleyic (hydromorphic) features
  6.2 Ferrods – others with little organic carbon in the spodic horizon
  6.3 Humods – others with little iron in some part of the spodic horizon
  6.4 Orthods – others with a spodic horizon containing aluminium, iron and organic carbon in which no one element dominates
7 **Alfisols** Soils with an argillic horizon and moderate to high base status
  7.1 Aqualfs – with gleyic (hydromorphic) features
  7.2 Boralfs – others of cold climates
  7.3 Ustalfs – others of warm, sub-humid or semi-arid climates
  7.4 Xeralfs – others with long dry periods
  7.5 Udalfs – others of humid climates
8 **Ultisols** Soils with an argillic horizon and low base status
  8.1 Aquults – with gleyic (hydromorphic) features
  8.2 Humults – other humus-rich soils of mid- or low latitudes
  8.3 Udults – others of humid climates
  8.4 Ustults – others of warm regions with high rainfall and pronounced dry seasons
  8.5 Xerults – others with long dry periods
9 **Oxisols** Soils with an oxic horizon or with plinthite within 30 cm depth
  9.1 Aquex – with gleyic (hydromorphic) features
  9.2 Torrox – others of arid climates
  9.3 Humox – other humus-rich soils of low base status
  9.4 Ustox – others with long dry periods
  9.5 Orthox – others of humid climates
10 **Histosols** Soils in organic materials (peat soils)
  *A Never saturated with water for more than a few days*
  10.1 Folists
  *B Saturated with water for six months or more*
  10.2 Fibrists – mainly composed of little decomposed plant remains
  10.3 Hermists – mainly composed of partly decomposed plant remains
  10.4 Saprists – mainly composed of almost completely decomposed plant remains.

Table 9.2 US Soil Taxonomy

| Major group | Group | Major group | Group |
|---|---|---|---|
| **1  Terrestrial raw soils** | 1.1  Raw sands | | 5.6  Brown alluvial soils |
| | 1.2  Raw alluvial soils | | 5.7  Argillic brown earths |
| | 1.3  Raw skeletal soils | | 5.8  Paleo-argillic brown earths |
| | 1.4  Raw earths | **6  Podsolic soils** | 6.1  Brown podsolic soils |
| | 1.5  Man-made raw soils | | 6.2  Humic cryptopodsols |
| **2  Raw gley soils** | 2.1  Raw sandy gley soils | | 6.3  Podsols |
| | 2.2  Unripened gley soils | | 6.4  Gley podsols |
| **3  Lithomorphic soils** | 3.1  Rankers | | 6.5  Stagnopodsols |
| | 3.2  Sand-rankers | **7  Surface-water gley soils** | 7.1  Stagnogley soils |
| | 3.3  Ranker-like alluvial soils | | 7.2  Stagnohumic gley soils |
| | 3.4  Rendzinas | **8  Groundwater gley soils** | 8.1  Alluvial gley soils |
| | 3.5  Pararendzinas | | 8.2  Sand gley soils |
| | 3.6  Sand-pararendzinas | | 8.3  Cambic gley soils |
| | 3.7  Rendzina-like alluvial soils | | 8.4  Argillic gley soils |
| **4  Pelosols** | 4.1  Calcareous pelosols | | 8.5  Humic-alluvial gley soils |
| | 4.2  Non-calcareous pelosols | | 8.6  Humic-sandy gley soils |
| | 4.3  Argillic pelosols | | 8.7  Humic gley soils |
| **5  Brown soils** | 5.1  Brown calcareous earths | **9  Man-made soils** | 9.1  Man-made humus soils |
| | 5.2  Brown calcareous sands | | 9.2  Disturbed soils |
| | 5.3  Brown calcareous alluvial soils | **10  Peat soils** | 10.1  Raw peat soils |
| | 5.4  Brown earths (sensu stricto) | | 10.2  Earthy peat soils |
| | 5.5  Brown sands | | |

Table 9.3  British System for classifying soils

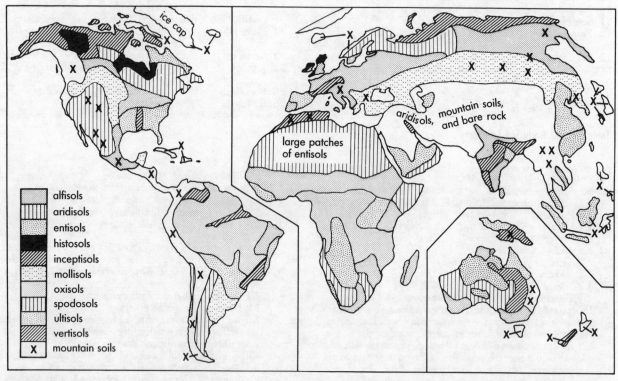

Fig. 9.2  World distribution of the ten soil orders

People disturb the soil so much that they could be seen as a soil-forming factor. They can affect **fertility** by increasing the potential **nutrient store**. They can add to the nutrient store with **organic manures** or manufactured inorganic **fertilisers**. **Soil structure** may be damaged by animal hooves, ploughing and vehicle weight. **Irrigation** and **drainage** also affect processes because of the role of water in pedogenesis. People also create artificial soils and activate erosion to cause total erosion.

We now look at a number of important concepts in pedology in rather more detail.

## TRANS-FORMATIONS

 *Types of transformation.*

Transformations are changes in the chemical composition of soil. Water is essential for weathering processes leading to soil formation.

- **Hydrolysis** is the breaking down of minerals by hydrogen so that they can be removed in solution – this is the most important transformation.
- **Chelation** – the removal of insoluble metals of iron and aluminium.

- **Hydration** – minerals absorb water and become more susceptible to chemical weathering – common in semi-arid areas.

- **Solution** – dissolving, for example calcium carbonate. Least soluble remain, for example silica. $CO_2$ in air within soil pores affects rate of solution.

- **Oxidation** – oxygen reacts particularly with the iron – ferrous iron converted to ferric iron – leads to reddish-brown colouring.

## PROCESSES OF PEDOGENESIS

### Upward transfers

Transfers may be **upwards**, i.e. **capillary rise** caused by evaporation loss on surface. If the water table is close to the surface, saline groundwater is drawn up and sodium and potassium salts are precipitated on the surface following evaporation – **salinisation** which is highly saline is called **solonisation**.

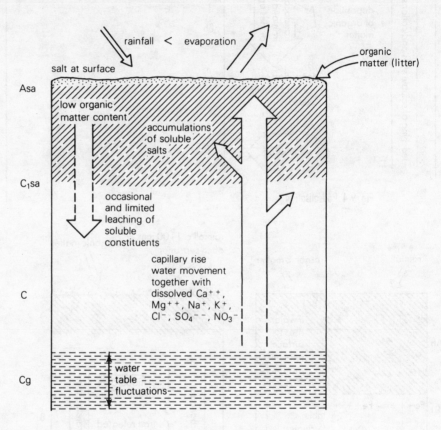

Fig. 9.3 Salinisation

### Downward transfers

**" Cases of downward transfer. "**

Movement *downwards* is most important:

- In **brown earths** (sol brun alfisols). Here **leaching** of soluble salts leads to decalcification.

- Where these soils are based on poorer parent materials and are freely drained, **lessivage** clay is mechanically translocated downwards, together with bases (sodium and potassium) and salts.

- In **podsolisation** (Fig. 9.4). Here the bases are rapidly leached. The iron and aluminium oxides are translocated to be deposited in B horizon. The trigger is possibly the presence of **mor humus** beneath slowly decomposing litter. The result is a characteristic podsol profile. Soil types where precipitation exceeds evaporation are often called **pedalfers**.

- In **ferralitisation** (Fig. 9.5). This is dominant in humid tropics (oxisols) where clay minerals break down and accumulate with oxides of aluminium and iron in the B horizons. These accumulations prevent root penetration laterite. Other terms used for tropical soils are **latosols** for the whole tropical area, **ferrisols** for those with strong red (iron) colouring.

   **Ferruginous soils** are those of the savannas but have less strong ferralitisation and leaching (Fig. 9.6).

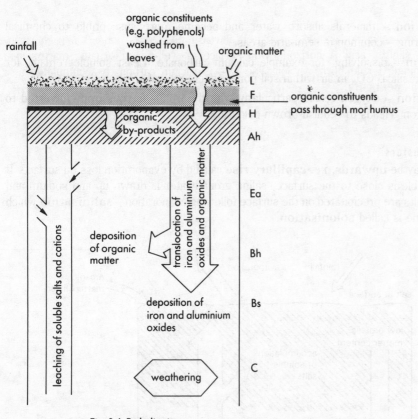

Fig. 9.4 Podsolisation

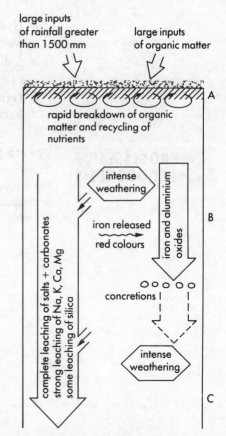

Fig. 9.5 Ferralitisation

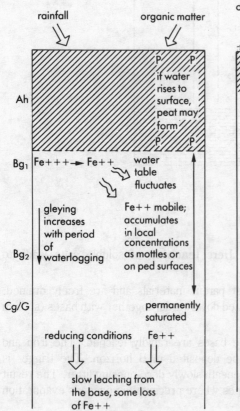

Fig. 9.6 Ferrugination

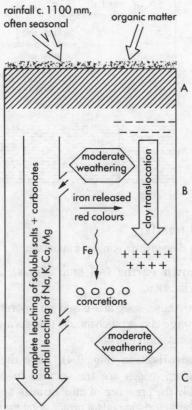

Fig. 9.7 Groundwater gley

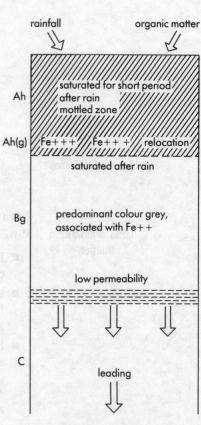

Fig. 9.8 Surface-water gley

■ In **gleization** or **gleying**. This is sometimes classed as a pedogenic process. It occurs where there is periodic or permanent waterlogging. If high groundwater level impedes water movement it leads to **groundwater gley** (Fig. 9.7).

If it is an impermeable layer restricting movement then it is a **surface-water gley** (Fig. 9.8).

■ In **calcification**. In steppe and prairie regions this takes place on **chernozems** (Fig. 9.9). Calcium carbonate is translocated and accumulates as nodules in B horizon. Leaves slightly acid A horizon. Mull surface of well-mixed organic matter in a free-draining fertile horizon. Soil type where calcium carbonate accumulates also known as **pedocals**.

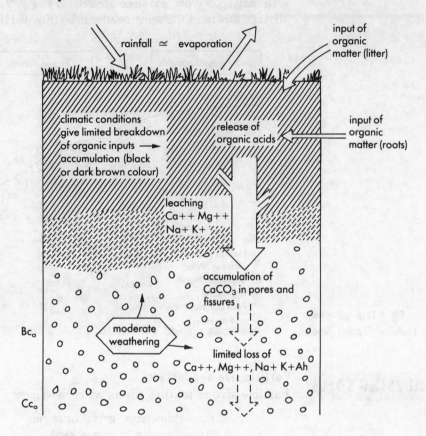

Fig. 9.9 Calcification and chernozem development

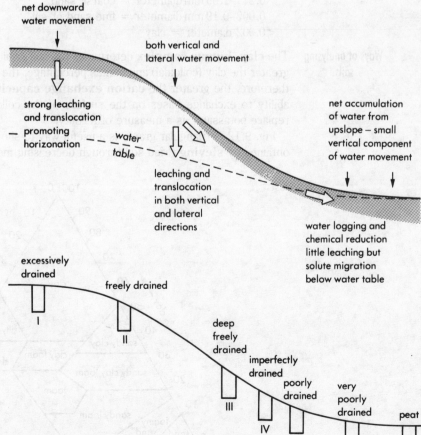

Fig. 9.10a) Soil slope and drainage relationships

Fig. 9.10b) How drainage then relates to catena classification I-VII

## PROCESSES ON SLOPES

> You can illustrate catenas from any slope – makes a good project.

The chain-like pattern of soils on a slope is known as a **catena**. These develop wherever the same topographic and climatic conditions occur. Soil profiles change with slope angle, drainage conditions, rock type.

Drainage conditions and slope are related as Fig. 9.10 a) and b) shows. A catena on the Mendips illustrates the same relationships (Fig. 9.11).

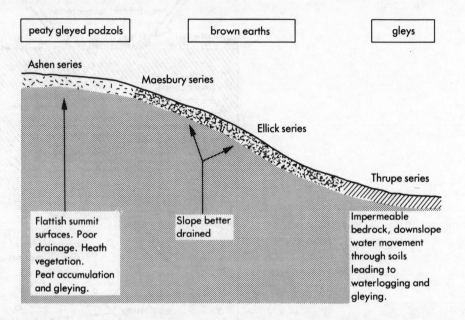

Fig. 9.11 A soil catena: Blackdown, Central Mendip

## ANALYSING SOILS

### Particle size and texture

**Particle size** or **texture** affects water movement.

$$\begin{aligned}
&>2 \text{ mm} &&\text{diameter} = \text{gravel or stones} \\
&0.2\ \ -1.99 \text{ mm} &&\text{diameter} = \text{coarse sand} \\
&0.002\text{--}0.19 \text{ mm} &&\text{diameter} = \text{fine sand or silt} \\
&<0.002 &&\text{diameter} = \text{clay}
\end{aligned}$$

> Ways of analysing soils.

The **clay–humus complex** determines the soil's ability to hold nutrients (**cations**). The greater the clay (colloidal) and humus percentage, the more the clay–humus complex and, therefore, the greater the **cation exchange capacity**. Cation exchange capacity is the ability to exchange bases on the surface of the colloids, e.g. hydrogen from water to replace potassium, is a measure of fertility.

Fig. 9.12, a triangular graph, is a useful way to show texture of a soil. The data can be obtained by **sieving** dried soil through decreasing meshes.

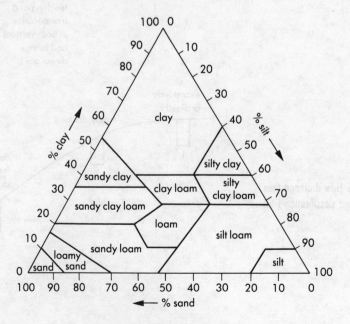

Fig. 9.12 Triangular graph of soil texture

## Structure

**Structure** is the most important property of soil as regards people's use of the soil. Particles bind together as **peds** which enable soil to stand up to erosive agents. The four types of structure are shown in Fig. 9.13.

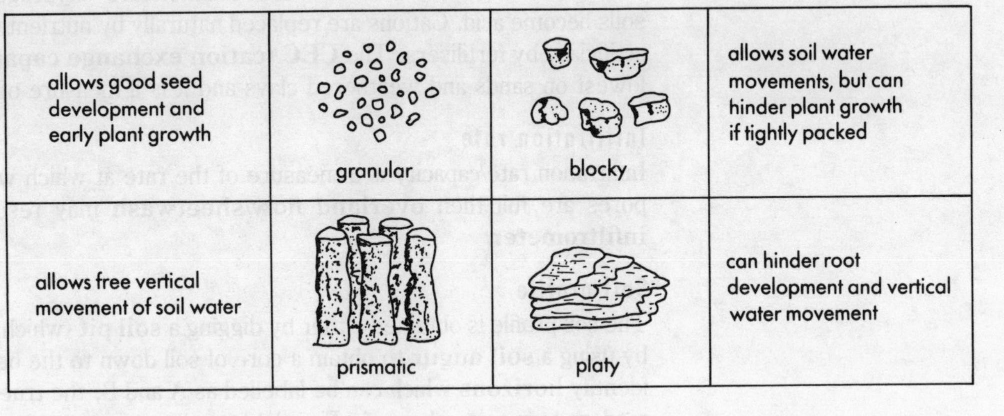

| | | | |
|---|---|---|---|
| allows good seed development and early plant growth | granular | blocky | allows soil water movements, but can hinder plant growth if tightly packed |
| allows free vertical movement of soil water | prismatic | platy | can hinder root development and vertical water movement |

**Fig. 9.13 Soil structures**

**Changes in structure** take place as the result of:

a) Ploughing, harrowing and rolling, which break down aggregates.

b) Deep ploughing in the tropics, which increases evaporation and therefore indigenous hoeing is more ecosystem-friendly. Excessive ploughing may break down structure and increase danger of erosion.

c) Heavy machinery which can cause formation of **ploughpan** due to compaction of clay particles.

c) **Poaching** (the effect of animals is similar).

## Texture

**Different textures.**

Texture is measured by rubbing soil between fingers.

1 Loose     non-coherent and crumbly when moist or dry;
2 Friable    when moist, soil crushes under gentle pressure but coheres if squeezed hard;
3 Firm      when moist, offers resistance to gentle pressure but crushes;
4 Hard      when dry, very difficult to break down;
5 Compact   firm and difficult to squeeze;
6 Plastic    when moist, can be moulded and rolled without sticking;
7 Sticky     plastic, but adheres to the hands.

## Colour

Colour indicates many properties of the soil; red-brown indicates ferric iron in well-drained soils (podsols), whereas blue-grey indicates ferrous iron compounds in waterlogged (gley soils). Chernozems tend to have a black appearance from humus, but black soil might be an infertile peaty soil. Mansell colour charts are used to check colour.

## Acidity (pH)

pH is the acidity of a soil or the hydrogen-ion concentration. It can easily be measured, using a kit. Acidic peats have a value of 4, neutral soils 7, and alkaline sands 10. pH influences nutrient supply which is vital for agricultural output. As alkalinity increases, iron, manganese and zinc nutrients are less available to plants. With acidity levels below 5, iron and manganese are soluble and are toxic to some plants. Therefore acid soils are **neutralised** by adding calcium carbonate ($CaCo_3$). Different crops need different acidity – some are **calciole** and like lime-rich soils and others **calciphobe** or **calcifuge** and cannot tolerate lime, for example plants on shell sand dunes at Bettyhill, Sutherland. pH meters which measure acidity electronically are also used and these are much more sensitive to variations in acidity.

## Capillary potential (pF)

pF or capillary potential is measured using a **tensiometer**. This indicates the energy involved in the retention of soil water and therefore the water-retaining potential of a soil. **Neutron probes** also measure moisture. The simplest method of measuring moisture is to take samples of soils in waterproof bags and weigh them before and after drying in an oven. This is expressed as a percentage of the oven dry weight.

### Base exchange or cation exchange

**Cations** are positively charged ions that move by electrolysis. They are held on colloids but are exchanged for other cations in soil moisture – hydrogen. If cations are not replaced, soils become acid. Cations are replaced naturally by nutrient cycling – decomposition – and artificially by fertilisers. The **CEC** (**cation exchange capacity**) is highest on humus and lowest on sands and weathered clays and it is a measure of nutrient reserves.

### Infiltration rate

Infiltration rate/capacity is a measure of the rate at which water enters a block of soil. If pores are full then **overland flow/sheetwash** may result. It is measured using an **infiltrometer**.

### Soil profile

The soil profile is obtained either by digging a **soil pit** (which would require permission) or by using a **soil augur** to obtain a core of soil down to the bedrock. You should be able to identify **horizons** which can be labelled as A and B, the true soil, and C, the subsoil. The modern terms are shown in Fig. 9.14.

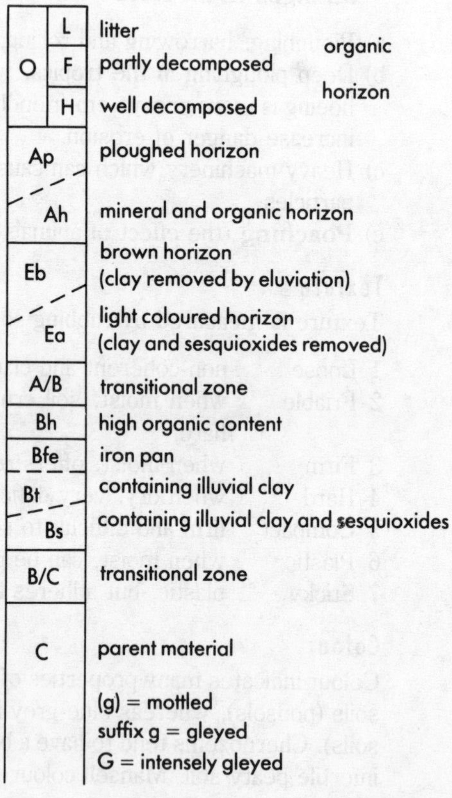

| | | |
|---|---|---|
| O | L | litter |
| | F | partly decomposed |
| | H | well decomposed |

organic horizon

| | |
|---|---|
| Ap | ploughed horizon |
| Ah | mineral and organic horizon |
| Eb | brown horizon (clay removed by eluviation) |
| Ea | light coloured horizon (clay and sesquioxides removed) |
| A/B | transitional zone |
| Bh | high organic content |
| Bfe | iron pan |
| Bt | containing illuvial clay |
| Bs | containing illuvial clay and sesquioxides |
| B/C | transitional zone |
| C | parent material |

(g) = mottled
suffix g = gleyed
G = intensely gleyed

**Fig. 9.14 Modern terms for soil horizons**

You can then measure particle size, structure, texture, colour, pH, in a series of pits, for example along a transect across different rock types up/down a C slope. The same measurements may be taken within a profile's horizons.

## AGRICULTURE AND SOIL MODIFICATION

Soils have been shown to be increasingly susceptible to damage from modern agricultural practices:

- **Structural damage** – when the acid content of clay soil is high, ploughing may damage the structure of the soil.
- **Ploughpan** compaction may occur where the base of the plough reaches. This can be combated by **zero-tillage**, sowing into holes without ploughing.
- **Compaction** by grazing animals, for example around waterholes and food supplies.
- **Wilting point** may be reached. A danger is the accumulation of salts which can reduce the productivity of crops, especially in semi-arid areas, for example Israel.

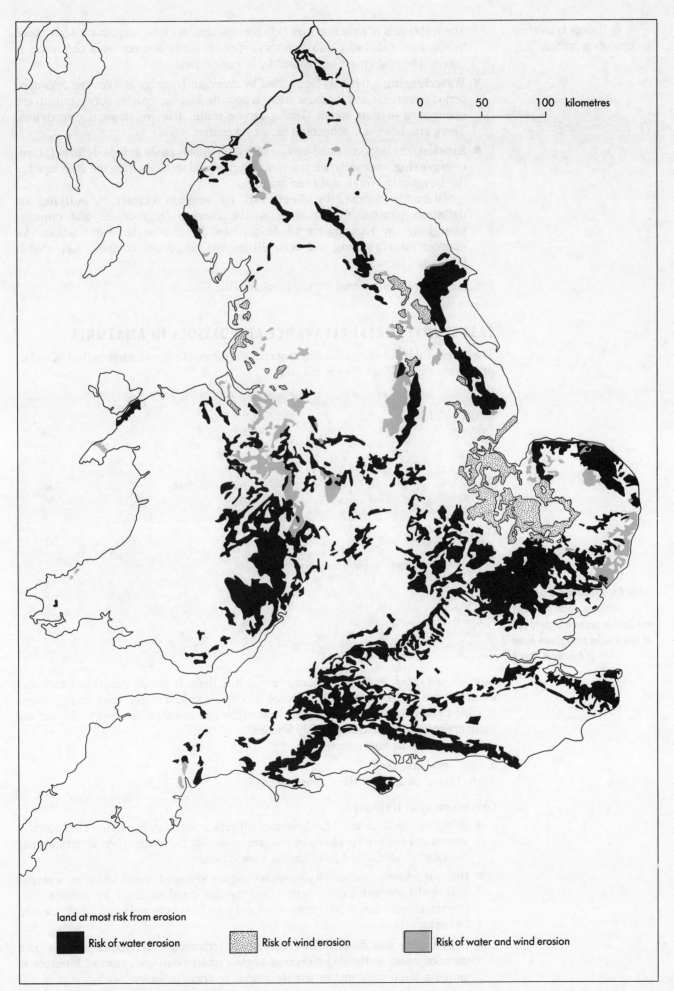

0      50      100 kilometres

**land at most risk from erosion**

■ Risk of water erosion        ⬚ Risk of wind erosion        ▨ Risk of water and wind erosion

Fig. 9.15 The potential for soil erosion in the UK

The techniques of irrigation may help prevent this. Such techniques vary from flood to drip feed and, today – hydroponics, spraying water and nutrients onto roots in gravel. The costs must be justified by increased yields.

- **Waterlogging** – this may be avoided by drainage. Drainage enables the cultivation period to increase and, because there is no waterlogging, enables spring growth and germination to start earlier. Drains can be **mole, tile** and major ditches/**dykes**. There are, however, dangers of increased erosion.

- **Erosion** may be accelerated by agriculture. Possible causes include **deforestation, overgrazing**, unresearched transfer of agricultural methods from the Developed to the Developing World, and even drainage.

  Soil may be removed by **sheetwash, rill erosion** (chapter 6), **gullying** and **deflation** (chapter 6). Erosion may be checked by rotations and **contour ploughing**, by reducing animal density, and by altering farming practices, for example **intercropping** and **zero-tillage** and hedgerow retention. Also shelter breaks and terracing.

Fig. 9.15 shows the potential for soil erosion in the UK.

## CASE STUDY: FOREST CLEARANCE AND OXISOLS IN AMAZONIA

The data in Table 9.4 shows the different characteristics of a brown earth in the UK and an Amazonian oxisol. What does it tell you?

**BROWN EARTH**

| Depth cm | % Org. matter | % Sand | % Silt | % Clay | pH | CEC |
|---|---|---|---|---|---|---|
| 0– 6 | 10.4 | 39 | 37 | 14 | 4.8 | 25 |
| 6 – 18 | 5.0 | 49 | 36 | 15 | 4.5 | 17 |
| 30– 64 | 0.3 | 30 | 6 | 64 | 5.8 | 36 |
| 64– 93 | 0.1 | 7 | 6 | 87 | 6.3 | 42 |
| 93–108 | 0.2 | 5 | 4 | 91 | 6.5 | 52 |

**OXISOL**

| Depth cm | % Org. matter | % Sand | % Silt | % Clay | pH | CEC |
|---|---|---|---|---|---|---|
| 0 – 8 | 3.4 | 10 | 15 | 75 | 3.4 | 18 |
| 8 – 18 | 0.6 | 11 | 11 | 78 | 3.7 | 8 |
| 18 – 50 | 0.5 | 7 | 8 | 85 | 4.2 | 4 |
| 50 – 90 | 0.3 | 7 | 4 | 89 | 4.5 | 3 |
| 90 –150 | 0.2 | 7 | 3 | 90 | 4.7 | 3 |
| 150–170 | 0.2 | 5 | 3 | 92 | 4.9 | 2 |

Table 9.4 Comparison of a brown earth soil under deciduous woodland in southern England and an oxisol under tropical rain forest in Amazonas, Brazil

Low pH and cation exchange capacity tell us that there is no soil reserve of nutrients except in the surface layer of the oxisol. Remove forest and you remove this store. Burning causes losses because the nutrients in the biomass store are lost to the soil and the nutrients in the ash can be easily leached.

Clearance can be undertaken by:

1 Slash and burn; or
2 Mechanical means, but three times as costly.

### Consequences of clearance

- With mechanical means – the bulldozer left ruts in which gullies form. The topsoil is uneven and removed in places by machinery, so that the fertile store of nutrients has been disturbed. Burned plots had no such variations.

- Heavy machinery causes soil to compact and dry less easily down to 15 cm, whereas slash and burn had no such effect and the soil dried as easily as under forest. Infiltration under forest 200 cm/hr, slash and burn 192 cm/hr and mechanically cleared 39 cm/hr.

- Soil moisture was similarly lowered by both techniques of clearance because both methods result in the destruction of organic matter during clearance. Moisture is stored in the organic matter and the matter restricts moisture loss.

- Burned plots yielded more even when fertilisers were added. Ash reduces acidity, although organic material was lower than under forest. Organic material was less on the bulldozed sites. On the whole, ash's contribution to crop yields declines with fertility of the soil. It also declines with time and more rapidly in low fertility soils.

# EXAMINATION QUESTIONS

### Question 1
Fig. 9.16 shows an idealised soil profile.

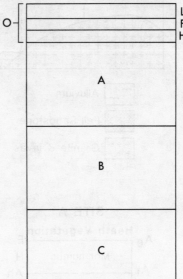

**Fig. 9.16**

a)  i)  Identify the horizons:
    A
    B                           *(2)*

  ii)  Identify the layers
    L
    H                           *(2)*

b)  i)  Under what conditions might Horizon A be thinner than Horizon B?    *(4)*

  ii)  Explain why the thickness of Horizon O will vary over a year.    *(2)*

c)  Outline the distinguishing characteristics of Horizon B in:
  1  a gley soil;
  2  a podsol.                                   *(6)*

d)  i)  What factors determine the structure of Horizon B in a soil?    *(3)*

  ii)  How might characteristics of Horizon B be altered by human activity?    *(3)*

e)  What factors determine the pH value of Horizon A in a soil?    *(3)*

                                                       (London, 1989)

### Question 2
a)  Define the following terms, which are all used in the study of soils:
   i)  anaerobic;
   ii)  leaching;
   iii)  the clay–humus complex;
   iv)  cation exchange                                   *(8)*

b)  Give an account of the processes by which organic matter is decomposed and incorporated into a soil, using examples of particular soils to illustrate your answer.    *(6)*

c)  With reference to any one soil type you have studied, explain how the nature of the clay–humus complex influences its natural fertility and potential for agricultural usage. State clearly the *name* of the soil chosen.    *(4)*

d)  Describe how the cation exchange capacity of a soil may be modified by human activity to upgrade and/or degrade a soil.    *(7)*

                                                       (JMB, 1987)

### Question 3

Fig. 9.17 shows a relief and soils transect across part of north-eastern England. Fig. 9.18 shows two soil profiles from the same area.

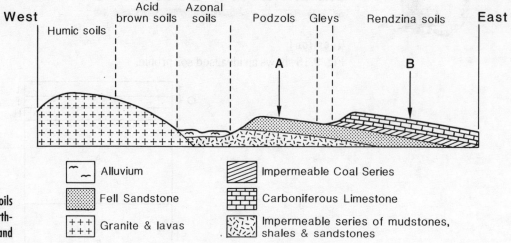

Fig. 9.17 Showing a relief and soils transect across a part of north-eastern England

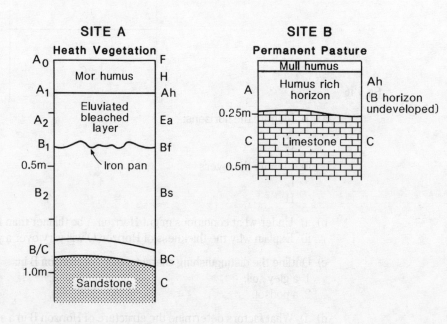

Fig. 9.18 Showing two soil profiles from the same area of north-eastern England

i) Discuss the influence of parent material in the development of the profiles shown in Figure 9.18. *(8)*

ii) Explain the importance of climate and parent material in the development of the humic peaty soils shown in the transect. *(9)*

iii) Explain why azonal soils and gley soils have developed in areas shown on the transect. *(8)*

(Oxford, 1989)

### Question 4

Study the diagram below which illustrates several types of soil structure.

a) Identify TWO of these types which are labelled A to D on the diagram.

A _____

B _____

C _____

D _____ *(3)*

b) Outline THREE ways in which structure may affect the physical properties of a soil. *(6)*

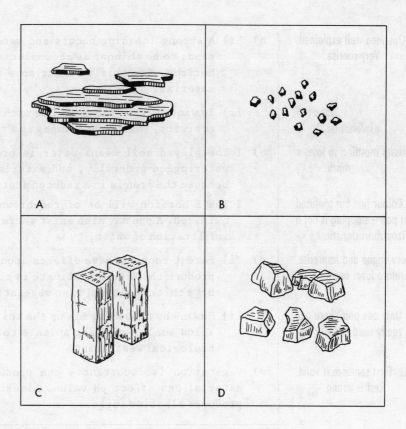

Fig. 9.19

c) With the aid of examples, suggest THREE ways in which agricultural practices may
modify soil structure.

*(6)*

(AEB, 1987)

## Question 5
Explain the role of vegetation in the process of soil formation.

# OUTLINE ANSWER

### Answer to Question 5
Although this is basically a soils question it does overlap both biogeography and soils. The
question states 'vegetation' and crop vegetation might be accepted, but do not over-
emphasise this.

Vegetation is the supply of nutrients – organic matter – from decay. Decay produces
humus at the top of the profile in the **humification layer**. The different types of humus –
**mor, mull** and **moder** – affect soil processes, e.g. decaying conifer needles release
chelatory compounds into the litter, and these are important in podsolisation.

Acidity is affected by and in turn affects soil fauna – more acidic = fewer earthworms.
Vegetation can affect the micro-climate and therefore the amount of evaporation,
evapo-transpiration and thus groundwater in the soil. Roots incorporate humus to varying
depths, e.g. chemozems, and by assisting in the weathering process they help to create
greater soil depth.

On the other hand people, by removing vegetation, arrest soil development and
accelerate the destruction of soil by erosion.

### STUDENT ANSWER WITH EXAMINER COMMENTS
### Question 1

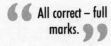
All correct – full
marks. 99

```
a)  i) eluviated horizon
       illuviated horizon

    ii) Leaf layer
        humus layer
```

b)
i) A strong leaching occurs and material is washed downwards. A tends to be thinner as accumulation of material occurs in the B horizon. For this to occur some water is needed to leach the material.

ii) In winter leaf litter will be much thinner because of less plant activity, therefore changes in a year.

c) 1 The gleyed soil means water is present. The soil will become waterlogged generally, and material will be blue-grey colour, because the iron is in a reduced state.

2 The B horizon will be of red-brown colour because the iron is oxidized. A pan may also exist where it becomes dry. This can stop infiltration of water.

d)
i) Parent rock can have effects upon the B horizon; clay tends to produce thick soils. Climate is also important. Tropical soils have thick horizons, also vegetation cover.

ii) Humans by ploughing mix up the soil. This breaks it down, it can allow small burrowing animals to enter and a larger amount of biological weathering.

e) Vegetation is important – can produce an acidic humus. Recent material can affect pH value. Limestone is an alkaline rock and produces alkaline soils.

CHAPTER

# ECONOMIC GEOGRAPHY

## GETTING STARTED

Economic Geography is one of the most diffuse sections of study at Advanced and AS Level. Almost all boards require you to study **agriculture**, although the emphasis does vary. For instance, London 16–19 places much greater stress on the environmental aspects of agricultural change, whereas London 210 has a more traditional approach via models and agricultural systems.

**Industrial geography** has traditionally been a focus of study because industry was, until the last 30 years, the dominant form of employment in Great Britain. In many syllabuses, industry and the exploitation of raw materials are covered together in the same section. Today more people are employed in service industries, and syllabuses recognise the importance of work in offices, shops and leisure.

**Communications, transport** and **trade** are not covered so fully and many centres choose to leave out one or all of these topics so as to ease the load.

One theme common to all studies of economic geography is the **role of government activity** in altering the patterns of economic activity – either directly by the law, or indirectly as an unforeseen consequence of its actions.

# ESSENTIAL PRINCIPLES

KEY CONCEPTS IN
ECONOMIC
GEOGRAPHY

- **The primary sector** is that part of the economy involved with the production, collection and use of natural resources. It includes agriculture, fishing, forestry and mining.
- **The secondary sector** is the sector of the economy which processes primary products and other products of the secondary sector into manufactured goods. It is, in other words, industry.
- **The tertiary sector** is the part of the economy involved with the distribution and retailing of manufactured goods and primary products, and includes a number of professional and personal services and public administration.
- **The quaternary sector** is that growing sector of the economy where high levels of skill and expertise are involved in the transmission and development of *ideas*. It is based on people and information and comprises research and development (R & D), financial services, a range of professional services and education.

## DEVELOPMENT STAGE MODEL

The four sectors we have outlined may be seen as part of the **development stage model** (Fig. 10.1) with its six stages:

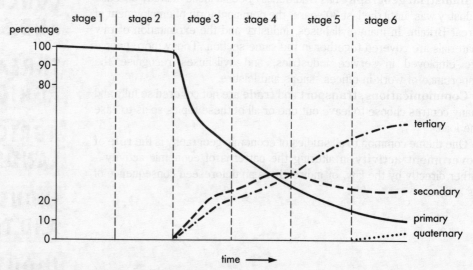

Fig. 10.1 The development stage model

❝ Be familiar with the development stage model. ❞

1 A self-sufficient primary sector dominates;
2 Primary sector dominates but specialisation exists within it and trade occurs;
3 Secondary sector develops with a small range of goods;
4 Secondary sector dominates with a wide range of goods demanded by increasingly wealthy society;
5 The tertiary sector grows to trade these products and begins to dominate the other sectors;
6 The emergence of a quaternary sector to develop still further a range of ideas and technologies.

AGRICULTURE:
CONCEPTS AND
THEORIES

Agriculture can be regarded as a **system** (for system definitions see chapter 6). There are climatic, soil and human inputs which vary from farming system to farming system. For example, the human **inputs** comprise the way in which farms are managed, the availability of technology, machinery and labour, the availability of seeds and fertilisers, and the support which comes from government. All the inputs combine within the agricultural system (the farms of an area) to produce products which are the **outputs** of the system to be processed and traded for the home or export markets.

Agriculture can also be regarded as an **ecosystem**. It is a **simplified food chain** in which, in the case of wheat, the inputs from solar energy, water and the soil are used to produce the crop. This is harvested without much decayed matter being returned to the soil other than some stubble. The food chain is more complex in the case of animal husbandry, with a more complex set of autotrophs consumed by the herbivores.

❝ Look at agriculture as a system, with inputs and outputs. ❞

Agriculture is limited by **climatic influences**: hence the similarity between climatic and agricultural regions. Both solar energy and precipitation influence agriculture. Although it is possible to enhance both by using glasshouses and irrigation, such practices hardly alter the basic patterns of agriculture. Soils are also determined in part by climate and these also influence agriculture. Only alluvial soils and loess are *not* climatically determined.

**Political intervention** in agriculture is increasingly important. Socialist ideologies have influenced the agricultural systems of Eastern Europe, while communist ideology has influenced China. Political groupings such as the European Community have policies which affect the nature and output of farms. National governments have policies which restrict output or control the market for a product, e.g. milk. Even decisions not primarily aimed at agriculture can have an impact on farming, such as the designation of a National Park or a Green Belt.

## VON THÜNEN'S LOCATIONAL THEORY

> Be aware of applications of the model.

Von Thünen's Theory of the 'Isolated State', although published in 1826, has become the basic theory of the location of agricultural production. He assumed: (a) one market, (b) one mode of transport, and (c) farmers supply that one market and respond to it. The theory was based on three principles:

1  The price obtained by a farmer declined with distance because the final price received for a product was the selling price minus the cost of transport.
2  The nearer the farm to the market the higher the returns, because the competition for the land raised its price and therefore forced farmers in such locations to make the most profitable, intensive use of the land.
3  **Economic rent** is the value of the crop minus the production and transport costs, and farmers would select those crops which gave the highest economic rent. The economic rent for each crop fell with distance from the market, which is how concentric rings for each crop were derived for crops whose economic rent diminishes at different rates.

Fig. 10.2 shows an application of the von Thunen model to the location of agricultural production around Sydney, Australia.

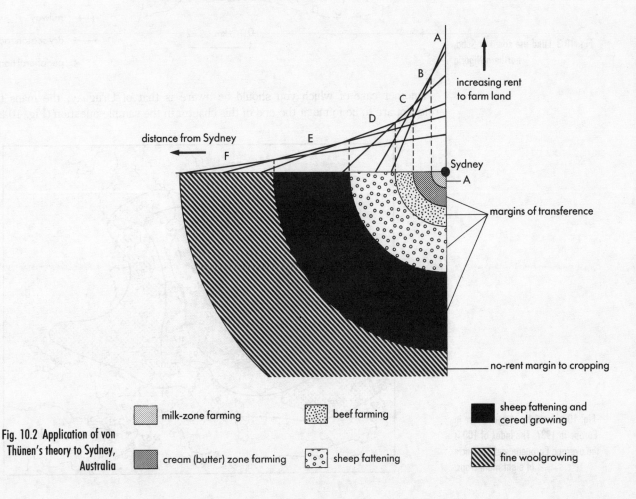

Fig. 10.2 Application of von Thünen's theory to Sydney, Australia

milk-zone farming   beef farming   sheep fattening and cereal growing

cream (butter) zone farming   sheep fattening   fine woolgrowing

Von Thünen did vary the theory to take account of a navigable river and a small city with its own trade area. Other variants of the theory will follow in the next sections.

Not all farmers are rational economic people who wish to maximise their profits, i.e. **maximisers**; many are **satisficers** who only change their cropping once they see a necessity to do so. Many choose to keep a horse for recreation rather than grow oil-seed rape because they gain greater **psychic value** from horse-riding than extra income.

## Von Thünen's Theory applied and evaluated

66 **Be familiar with this model.** 99

Make sure that you are aware of *examples* of the application of the model at all scales from the farm to the continental scale, in a variety of locations. At the farm scale there is scope for a project to see if the most intense activity takes place nearest the farm. In the Developing World several studies have been made which show the application of the model to simple agricultural societies, such as those of the savanna region in northern Nigeria (Fig. 10.3).

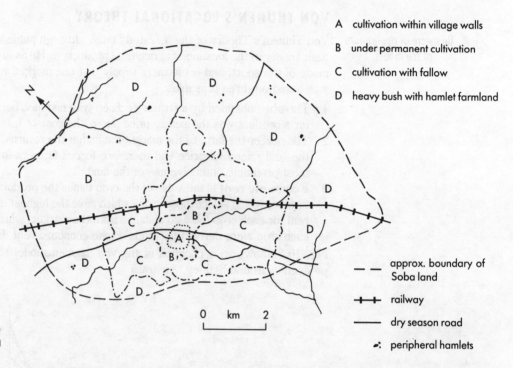

A   cultivation within village walls

B   under permanent cultivation

C   cultivation with fallow

D   heavy bush with hamlet farmland

- - -   approx. boundary of Soba land

+++   railway

—   dry season road

∴   peripheral hamlets

0   km   2

Fig. 10.3 Land-use zones at Soba, northern Nigeria

Another case of which you should be aware is that of Uruguay; the maps that are relevant are to be found at the end of this chapter in the sample question (Fig. 10.23 a–c).

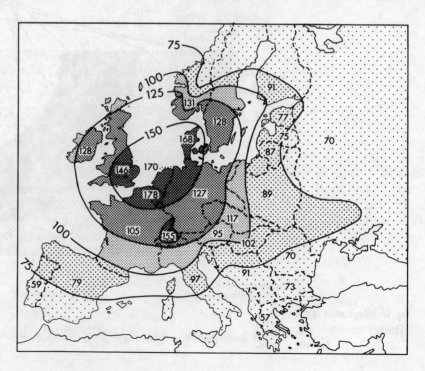

Fig. 10.4 Intensity of agriculture in Europe in 1937. The index of 100 is the average European yield per acre of eight main crops.

A further study of New South Wales, Australia, showed how the terrain modified the anticipated pattern of uniform transport costs to give a north–south arrangement of uses related to climate and soils. The model was further distorted to take account of the coastal land uses.

A final case of the application of the model at the continental scale was the study of the intensity of European agriculture which produced a series of concentric zones focused upon north-west Europe (Fig. 10.4).

Since 1826 farming methods have changed and the modifying influences have changed, as von Thünen actually stated. The changed circumstances can be summarised:

- There is more than one market which will distort patterns.

- Transport methods and accessibility have changed so that produce can be grown where the physical conditions are more beneficial – refridgerated transport and air transport are major changes.

- New crop strains have enabled crops to be grown in new areas, e.g. maize in the UK.

- Grain farming is now practised much further from markets. However, by growing grain more intensively with higher labour and fertiliser inputs it is possible to continue production in north-west Europe.

- Political and social influences on agriculture are more important today. Nevertheless von Thünen did note that an educated workforce would increase productivity and that imports would alter patterns.

- The surface is **isotropic** and assumes the same physical conditions everywhere.

### Sinclair's modification of von Thünen

Sinclair examined the impact of urbanisation on agriculture on the urban fringe. He hypothesised that, contrary to the assumptions of von Thünen, land close to the city margin had a low value because of the damage done to crops by vandalism and the holding of land speculatively in the anticipation of development. Very often low intensity uses creep in, such as the keeping of horses and other pets. As a result, the value of land for agriculture near the city margin is lower than in areas more removed from the city. Efficient transport also means that intensive agriculture may be located further away from the market.

## AGRICULTURAL SYSTEMS

## INTENSIVE AGRICULTURE

Intensive agriculture is agriculture which has high levels of capital, labour and fertiliser inputs and high levels of output for each hectare of land. The most intensive forms of agriculture, such as market gardening, horticulture and glasshouse crops, take place in relatively small areas (Fig. 10.5).

The reasons for the patterns are to do with:

a) Market, e.g. the Lea Valley and North Kent for London;
b) Physical conditions with more solar radiation and higher temperatures;
c) Good soils, e.g. brickearth, West Sussex plain;
d) The historical acquisition of expertise in a crop, e.g. Guernsey tomatoes.

*Types of intensive agriculture.*

Freezing and canning has also resulted in the development of intensively farmed areas of supply for such industries. Sometimes these developed to freeze fish, e.g. Lowestoft and Grimsby, and then vegetable (pea) freezing developed alongside. The technology demands that the crops are frozen within a short period of harvest, so e.g. the pea fields cluster close to the freezing factories.

New **bio-technologies** such as **hydroponics** enable plants to be grown in, for example, polystyrene, the roots being fed with the correct nutrients for optimum growth to supply the needs of the supermarket chains. Control of light and temperature is used to produce e.g. flowers for Christmas which naturally flower in the autumn and spring.

Transport enables intensive farming to go further afield: in the USA, **truck farming** enables produce from California and Florida to be moved by refrigerated lorry or by air to the markets of the north-east.

There are other forms of intensive agriculture: (1) pick-your-own farms which have remained close to markets (why?), (2) battery farming of poultry, and (3) intensive dairying and meat production, including intensive deer farming. **Irrigation agriculture** is another intensive system.

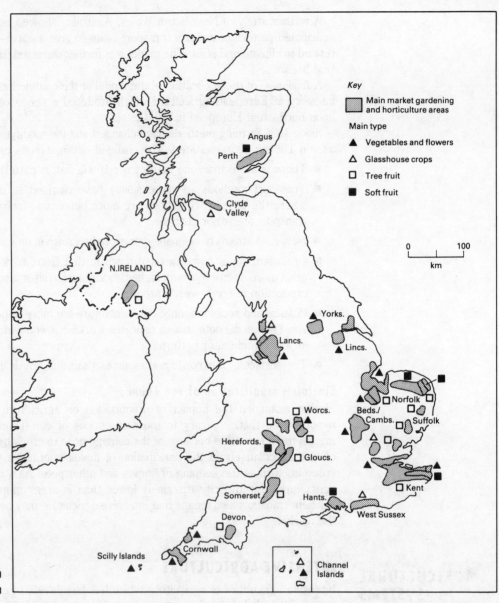

Fig. 10.5 Intensive agriculture in Great Britain

## EXTENSIVE AGRICULTURE

Extensive agriculture is characterised by low levels of labour and capital input per hectare of land and a low level of output per hectare. In contrast to intensive agriculture it is characterised by:

1  Large units of land holding;
2  High levels of mechanisation;
3  Very little labour, but a high yield per worker.

The term is a relative one because the 'barley barons' of Hampshire are regarded as extensive farmers in England but, compared with cereal-growers in North America their yields seem like intensive agriculture. Extensive agriculture can also be practised in **marginal areas**, where revenues are close to costs, or where physical conditions are harsh, e.g. the uplands of central Wales. Some people also regard **shifting cultivation**, where a semi-nomadic group cultivate a small area of land until it is exhausted, and then abandon it and move on to a new area, as a form of extensive agriculture found in parts of Africa and South-East Asia.

## PLANTATION AGRICULTURE

Plantation agriculture is a commercial farming system associated mainly with the colonial period in the tropics. It involved the growing of one crop for profit, with high initial capital input, low-cost intensive labour input, and highly profitable output of specialist crops such as rubber in Malaya (Malaysia) and tea in Ceylon (Sri Lanka). Such crop patterns were to increase dependency (chapter 12) after independence, and so most of the countries have diversified their agriculture/production. Consequently plantation agriculture has declined.

## FARM SYSTEMS AND CLIMATIC ZONES

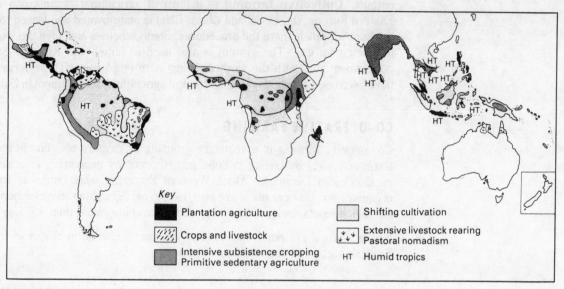

Key

▪ Plantation agriculture   ▨ Shifting cultivation

▨ Crops and livestock   ↓↓ Extensive livestock rearing Pastoral nomadism

▨ Intensive subsistence cropping Primitive sedentary agriculture   HT Humid tropics

**Fig. 10.6** Land-use patterns in the humid tropics

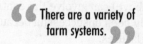

❝ There are a variety of farm systems. ❞

**Fig. 10.7** Types of temperate farming systems

It is also possible to classify agricultural systems according to the climatic, social and economic organisation dominating in different regions. Tropical systems (Fig. 10.6) include shifting cultivation and plantations but also include other systems such as **pastoralism** which can either be **nomadic** (the movement of people and animals in response to climate and food supplies) or more **sedentary**, an example of the latter being the 'hamburger farms', such as those being developed by clearing the rain forests in Rondonia, Brazil, to provide cheap beef for the United States markets. A further tropical system is **intensive subsistence** and **export cropping** based on rice cultivation which dominates South-East Asia in particular (see the 'Green Revolution', below).

Temperate farming systems (Fig. 10.7) include commercial grain farming, livestock, ranching and nomadism, dairying, mixed farming, including the unusual mix of the

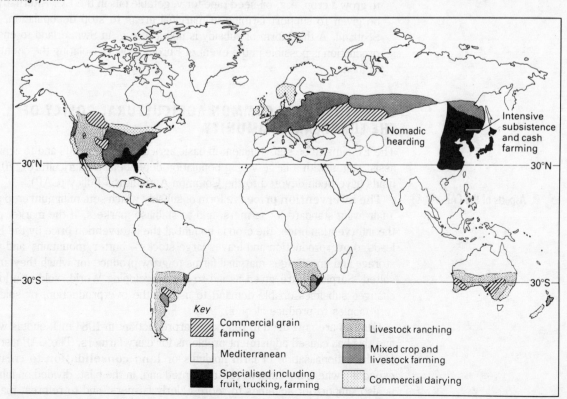

Nomadic hearding

Intensive subsistence and cash farming

30°N

30°S

Key

▨ Commercial grain farming   ▨ Livestock ranching

☐ Mediterranean   ▨ Mixed crop and livestock farming

▪ Specialised including fruit, trucking, farming   ▨ Commercial dairying

Mediterranean farming areas – olives, vines, intensive vegetables and some pastoralism – and areas of specialised farming. China is an exception to these patterns.

## SOCIALIST FARMING SYSTEMS

Socialist farming systems are distinguished by their organisation and not inputs and outputs. **Collective farming** is a form of agricultural organisation characteristic of Eastern Europe, the USSR and China. Land is state-owned and leased to a group of farm workers who then share the procedures. Some schemes also give the workers some land for their own use. The system is also used in Israel: the Kibbutz. **State farms** are government run, with the workers being state employees. These forms of organisation have been used and modified to aid tropical agricultural development in Cuba and Tanzania.

## CO-OPERATIVE FARMING

Co-operative farming is a voluntary grouping of farmers to gain cheaper inputs (seed, fertilisers, etc.) by buying in bulk, and to profit by economies of scale in output (e.g. creameries in Denmark). Much Western European wine output is from co-operative organisations. Co-operatives are also used to aid agricultural development in the tropics – for capital inputs such as machinery and to promote marketing.

No agricultural system is fixed. They constantly change in response to environmental, political, social and economic conditions.

## GOVERNMENT INTERVENTION IN AGRICULTURAL SYSTEMS

The basis of intervention is to ensure that the population is fed and, as far as possible, fed from within the country/economic bloc. Obviously the people are voters, and governments will be aware of electoral implications.

**Socialist intervention** (see Socialist farming systems, above) can involve the control of the output through targets in state plans. This has generally been unsuccessful because it does not take account of extreme climatic events or crop failures.

**State intervention** in capitalist countries takes the form of:

a) **Quotas** – a specified maximum production of a crop to restrain production – dairy produce in the European Community (EC) has been restricted like this.

b) **Subsidies** – extra cash is given to farmers, above the market price, to encourage them to grow a crop, e.g. oil-seed rape for vegetable oils in the EC. Alternatively subsidy can be given to support farming in marginal areas to stop depopulation, e.g. crofting in Scotland. A third form of subsidy is political, e.g. in Switzerland to ensure agricultural production is possible in the event of European wars isolating the country from imports of food.

## CASE STUDY: THE COMMON AGRICULTURAL POLICY OF THE EUROPEAN COMMUNITY

The EC aims to be self-sufficient in basic agricultural foodstuffs and to achieve this (and to assist states with a large voting population dependent on agriculture), 70 per cent of EC funds have been devoted to the Common Agricultural Policy (CAP).

**66 Aspects of the CAP. 99**

**The intervention price** is a form of subsidy which sets minimum crop prices to ensure a fair living standard for farmers, and to stabilise markets. If the market price falls below the intervention price, the crop is bought at the intervention price by the EC. The system leads to overproduction and leaves large stocks – butter 'mountains' and wine 'lakes' – in storage and encourages marginal farms to grow produce for which they are not physically suited. Surpluses are given as aid to the Developing World, sold at a loss, stored until changed subsidies enable demand to use up the overproduction, or sold for other uses (e.g. apples to produce alcohol).

**Quotas** are now used to restrict overproduction: in 1984 milk quotas were introduced, and this has caused adjustment problems for dairy farmers. The CAP also provides funds for the rationalisation of farm holdings or **land consolidation** to create bigger units, especially where holdings were fragmented and, in the past, divided on inheritance. There is also support for schemes to retire elderly farmers, and to reforest marginal land.

Finally, with the help of national governments, there is a **set aside** policy which is voluntary in the UK. Farmers are paid to take at least 20 per cent out of the farm area growing certain arable crops. The land can be forested or used for permitted non-agricultural use. This should also reduce surpluses.

### Land consolidation

Land consolidation, often known by its French term *remembrement*, is a government policy, also found in the Netherlands, West Germany, Switzerland and Austria, which is designed to amalgamate farm plots into compact holdings in which capital inputs can increase and labour inputs decrease, with consequent rises in output. It often involves building new farm buildings, dispersing the population from nucleated villages (chapter 11), besides building new roads.

## THE GREEN REVOLUTION

The 'Green Revolution' is a term given to the introduction of new, more productive agricultural techniques in the Developing World. It began with the development of high-yielding strains of wheat in Mexico and rice (IR8) in the Philippines, and could be seen as a response to population pressure of the kind outlined by Boserup.

The crops are more tolerant of marginal conditions, grow quickly – so permitting **double-cropping** of rice – but they depend on large inputs of fertilisers. Output has risen dramatically but it has merely kept pace with population growth. The Green Revolution has benefited the large land-owners who have the capital to purchase the complete package, but the small peasant farmer doesn't have the capital and is disadvantaged by exclusion or ignorance of how to apply the revolution. Government schemes such as co-operatives (see above) do help, but many, unable to cope, drift to the cities in search of work.

**THREATS TO AGRICULTURE**

## SOIL EROSION

Soil erosion (see also chapters 4 and 9) can be produced by:

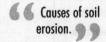 Causes of soil erosion.

a) **Overgrazing** – where the land is being asked to provide for too many animals – destroys vegetation, especially in areas of seasonal rain where grasses are eaten down and trampled, especially round water holes.
b) **Ploughing**, which leaves the surface bare and exposed to **sheetwash** and **rill erosion**.
c) **Stubble-burning**, which also exposes soil.

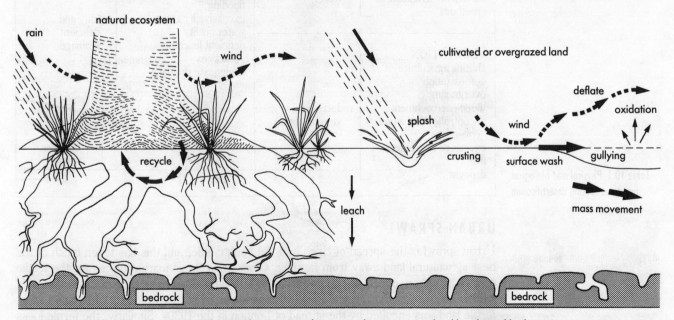

Fig. 10.8  The possible accelerating erosion mechanisms that can occur when a natural ecosystem is replaced by cultivated land

There are techniques to reduce erosion: rotation, contour ploughing, cover crops, windbreaks, irrigation and better watering for animals. These, together with changed social practices such as ceasing to use animals as a sign of wealth, can all contribute to **soil conservation**.

## SALINISATION

Salinisation is a natural process in arid soils (chapter 9). Where there is irrigation of crops about half of the water is lost through evaporation and evapo-transpiration. As a result of the former, sodium and potassium salts crystallise on the soil surface. Excessive salt accumulation caused by poor irrigation practice will reduce crop yields, e.g. in Israel cotton yields have declined where salt accumulations in the soil have built up after decades of irrigation. To solve the problem salts must be **leached** back to the river system; this takes more water and is best attempted if there is a dormant season.

## DESERTIFICATION

Desertification is the spread of desert conditions into the margins of deserts, as has occurred over the past two decades in the Sahelian zone of Africa. The cause is partly to do with global climatic change, but human activities such as overgrazing and the destruction of woodland and scrub for firewood also contribute to the process, which is enlarging deserts by 3 per cent every 25 years. Table 10.1 outlines a number of physical and biological processes causing desertification.

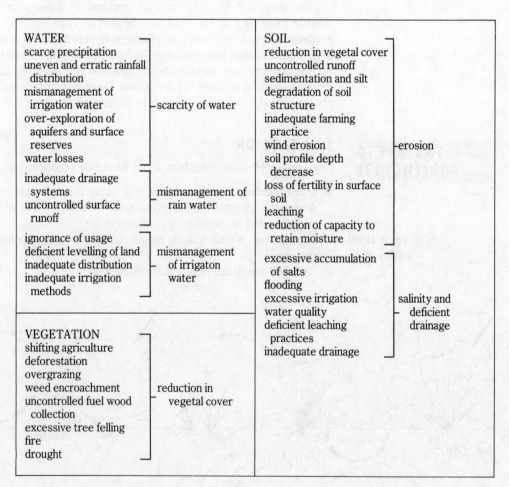

**WATER**
scarce precipitation
uneven and erratic rainfall distribution
mismanagement of irrigation water
over-exploration of aquifers and surface reserves
water losses
— scarcity of water

inadequate drainage systems
uncontrolled surface runoff
— mismanagement of rain water

ignorance of usage
deficient levelling of land
inadequate distribution
inadequate irrigation methods
— mismanagement of irrigaton water

**VEGETATION**
shifting agriculture
deforestation
overgrazing
weed encroachment
uncontrolled fuel wood collection
excessive tree felling
fire
drought
— reduction in vegetal cover

**SOIL**
reduction in vegetal cover
uncontrolled runoff
sedimentation and silt
degradation of soil structure
inadequate farming practice
wind erosion
soil profile depth decrease
loss of fertility in surface soil
leaching
reduction of capacity to retain moisture
— erosion

excessive accumulation of salts
flooding
excessive irrigation
water quality
deficient leaching practices
inadequate drainage
— salinity and deficient drainage

Table 10.1 Physical and biological processes causing desertification

## URBAN SPRAWL

Urban sprawl is the spread of cities into the countryside, and this has taken much of the best agricultural land away from farming. Cities grew up as market places for the fertile areas which they served, and therefore many cities are situated in some of the most fertile lands – the alluvial and terrace gravel soils of lower river valleys. Some of Britain's most fertile land was engulfed by the spread of London in the 1930s. Similarly, the fertile loess (brick-earth) soils of West Sussex were covered by the sprawl of the south coast retirement towns between 1930 and 1970. Heathrow Airport has covered up another area once important for market gardening. Urban phenomena can affect the local climate of the surrounding area, e.g. dust and haze and convection storms induced by the urban area. Urban drainage can affect local stream flow and bring water pollution to streams used by farmers, as well as altering the water table.

## AGRICULTURE AND LANDSCAPE CHANGE

Agriculture affects the landscape through:

a) Hedgerow removal;
b) Woodland clearance;
c) New buildings for machinery and animals, plus silos;
d) New crops introducing new colour – e.g. bright yellow of rape;
e) Pesticides and herbicides altering ecosystems;
f) Stubble-burning after harvest of cereals;
g) Polluting streams with slurry;
h) Using larger machinery which can **poach** soil and lead to erosion;
i) Nitrates being washed into streams causing eutrophication.

Why are these effects taking place?

1  Pressure for profit;
2  Need to provide food to suit tastes;
3  Government subsidies.

Why are people resisting the effects of modern farming?

1  The landscape is a tourist resource – it is part of national heritage;
2  Scientific research is showing the incremental nature of change on the environment and its ecosystems;
3  Greater public awareness as a result of pressure groups' activities.

## PROJECTS

- The next stage of an investigation into agriculture's effects could be an examination of landscape change in your area over the past 25 years using photographs, press reports and pressure group reports.
- Test von Thünen's theory in two contrasting areas or two contrasting farms.
- An investigation into soil loss caused by winter wheat growing.
- Agriculture on the urban fringe: is Sinclair correct?
- Land use, soils and geology in an area.
- The effects of set-aside: (you would need farming community contacts for this).
- Diffusion of the use of sewage sludge or any other agricultural innovation in an area.

## INDUSTRY: CONCEPTS AND THEORIES

### Weber's model of industrial location (1909)

This involves the least cost approach using the **material index**. If transport costs are identical for raw materials and finished products, then the location of manufacture would be equidistant between the raw materials and the market. If the weight of the raw materials is greater than that of the finished product, then location will be at or near the sources of the raw materials; this is a **weight losing industry**, such as iron and steel. Conversely if the industry is **weight gaining**, it will be sited nearer to the market, e.g. oil refining. The other considerations are **labour costs** and **agglomeration economies**, i.e. the savings made by locating close to specialist services and to like firms needing similar skills. **Deglomeration** forces can have the opposite effect and force companies out of an area, owing to factors such as congestion, skills shortages, and so on.

Weber used graphical means to show how the least cost situation was established (Fig. 10.9). **Isotims** are lines of equal cost of transport from a raw material source and from a market. **Isodapanes** are lines joining all points with the same total transport costs. They form **cost contours** around a least cost location. Weber assumed an **isotropic** surface: i.e. that the same conditions existed everywhere on the plain.

Weber's theory is merely a starting-point: it is now rather dated since conditions have changed and transport costs are not always the most important; it also emphasised costs and paid little attention to revenue. Political factors have now also come into play as have the structure of the firm and various behavioural factors.

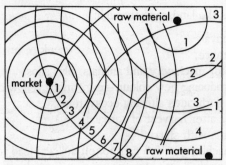
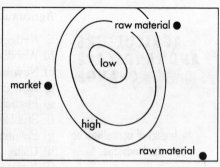

**Fig. 10.9 Isotims and isodapanes**     isotims                    isodapanes

### Lösch's market area theory (1954)

This emphasised the role of demand from the market which would maximise revenue; revenue would fall with increasing distance from the market (Fig. 10.10). This could lead to **agglomeration** or it could lead to dispersal as each company sought its own market. Lösch was slightly more realistic, but his theory still assumed perfect knowledge of revenues. It is most valid for brewing and baking.

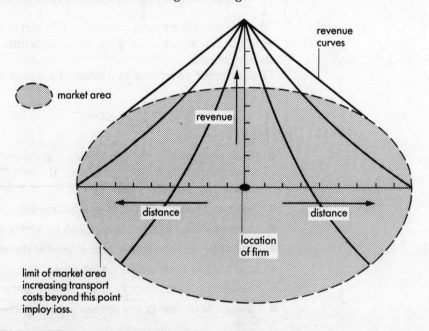

**Fig. 10.10 Lösch's revenue cone**

### Smith's spatial margins model or space cost curve (1966)

This plots both revenue and costs against distance to show that profits can be obtained in an area. For instance, the choice of locations in the area Y–Y$^1$ (Fig. 10.11) will all result in profit for the entrepreneur. It recognises the concept of **satisficers** in industrial location and it acknowledges that locational knowledge is not perfect. Still the theory concentrates on individual companies and decision-makers.

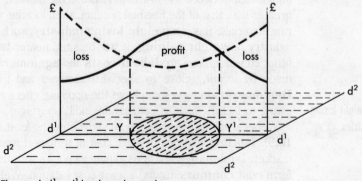

The graph d$^1$ – d$^1$ has been located
across an isotropic plain d$^2$ – d$^2$.

**Fig. 10.11 Smith's spatial margins model**

| | | |
|---|---|---|
| £ | e.g. revenue or cost | total cost curve | area of profit |
| total revenue curve | spatial margin | area of loss |

### Hoover's theory (1948)

This attempted to vary Weber by showing that transport costs were stepped, and decreased per unit of distance. He also noted **break of bulk points** or **trans-shipment points** as sites of lower transport costs.

### Hotelling's model (1928)

Fig. 10.12 shows how demand influences location in a linear market – namely ice-cream sellers on a beach. Stage 1 shows the socially optimum location, i.e. sharing the market (A); but to increase sales, firm X moves closer to Y to extend its market (B). The effect is that Y moves close to X to achieve **competitive equilibrium** (C).

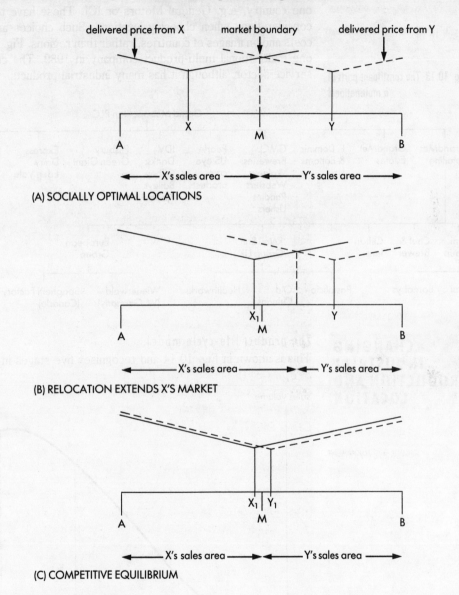

(A) SOCIALLY OPTIMAL LOCATIONS

(B) RELOCATION EXTENDS X'S MARKET

**Fig. 10.12 Hotelling's model**

(C) COMPETITIVE EQUILIBRIUM

### Behavioural approaches

These were developed in the 1960s to show that boards of directors have to make corporate decisions on the basis of imperfect knowledge of all the factors. The decision-makers will be influenced by a range of social, economic, political and business considerations. Studies of firms relocating their plants, and expanding in order to establish branch plants, have shown how decision-makers go for the known and familiar. Such decision-makers often have dated stereotypes of areas and seek to rationalise the irrational: the real reason for the choice may be that it is the managing director's birthplace! This approach focuses on the decision-makers themselves and ignores other influences.

### Structure of industry approaches

Most theory has tended to assume single-product firms. Today most companies are **multi-product enterprises** receiving raw materials from a host of sources and selling to

**Firms often have many products.**

many markets. Firms are owned by shareholders and often have share-owning workers interested in the firm's profitability. Many firms have more than one plant, i.e. they are **multi-plant**, with a main factory and **branch plants**; possibly they have a separate location for their office headquarters, with research and development in yet another location. Companies have tended to see some areas as providing basically **branch plant economies**, e.g. the electronics industry in Scotland, whereas industrial innovation, research and development is confined to other regions, e.g. M4 and M11 corridors and south Hampshire.

**Multi-national companies**, sometimes called **transnational corporations**, manufacture goods or provide services in several countries while being administered from one country, e.g. General Motors or ICI. These have to take into account even more considerations when choosing locations. Such choices are often based on comparative costs and on images of countries, rather than regions. Fig. 10.13 shows the organisation of one multinational multi-product company in 1989. The company operates mainly in the service sector, although it has many industrial products.

**Fig. 10.13 The constituent parts of a multinational**

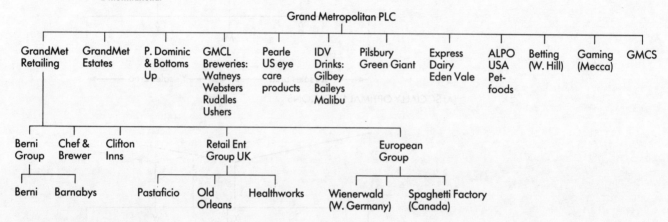

**CHANGING INDUSTRIAL PRODUCTION AND LOCATION**

### The product life-cycle model

This is shown in Fig. 10.14 and recognises five stages in the lifetime of a product:

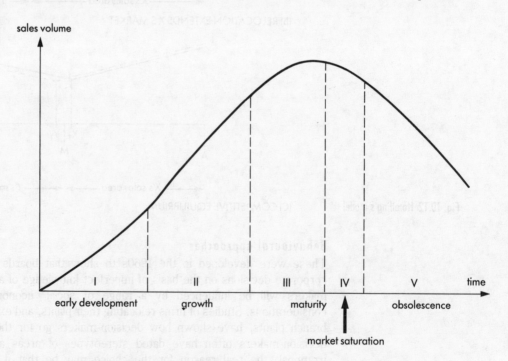

Fig. 10.14 The product life-cycle

**Stages of the product life cycle.**

1  The introductory phase – innovation;
2  The phase of growth – increased market penetration;
3  The mature phase – increased sales but also competition;
4  Market saturation;
5  Obsolescence – declining sales and a new product – and this has happened for many electronics products, e.g. CD replaces the hifi unit. The concept is important because

new producers don't wish to associate with old technology images and often want new locations.

## Rationalisation

This is the reorganisation into fewer factories to achieve greater efficiency and profit. It can follow a merger or changes in revenue, or increased competition from overseas. British Steel has rationalised its production from a whole series of plants at the old locations near sources of raw materials and break of bulk points, down to a few large-scale integrated plants e.g. Redcar. Uneconomic inland sites such as Corby have closed, and altogether a quarter of a million jobs have been lost in the steel-making towns.

## Deindustrialisation

This is the gradual decline in the contribution of industry to the economy of a region or a country. It is often seen as the decline in the number of manufacturing jobs. The reasons for it include:

> **If you live in a deindustrialised area, use that as your case.**

1   Cheaper products imported/manufactured elsewhere;
2   Products at the end of their life-cycle;
3   Production shifted out of the region/country to low-cost labour areas elsewhere;
4   Tax levels reducing company profits.

Deindustrialisation affects older industries/areas most, e.g. steel, textiles, shipbuilding. Losses of jobs in industry may be partly offset by the growth of jobs in tertiary and quaternary sectors. But often those who have lost jobs have got the wrong skills for the new jobs and anyway the new jobs are often not in the same region. Areas affected by deindustrialisation include South Wales, Lorraine, and the Ruhr coalfield.

## Industrial inertia or geographical inertia

This is the tendency for an industry to remain in an area despite more favourable location factors existing elsewhere. Very often the capital invested in the site is the main deterrent to movement. The steel industry remaining in Dortmund is a good case, because Duisburg on the Rhine has a better location – though even Duisburg is not ideal these days.

## Newly industrialising regions

These are areas where companies are locating their innovative activities. In the UK the M4 corridor from Heathrow Airport west to Reading, Swindon and Bristol is one such region. Some 60 per cent of the new high-technology firms have been established here in the past 20 years. This corridor is wider nearer London in the 'Golden Triangle': Reading–Basingstoke–Guildford–Heathrow. Other centres of innovation, such as a university with its research emphasis, also pull new firms into an area, e.g. MIT at Boston, Massachusetts, Stanford and the nearby Silicon Valley in California and the Cambridge Science Park established by Trinity College. Major research organisations of government also act as a magnet, e.g. Aldermaston (weapons), Farnborough (aircraft), Bracknell (meteorology) and Harwell (nuclear) are all along the M4 corridor. Such factors all lead to the **spatial division of labour** in the UK.

## Newly industrialising countries (NICs)

These include Brazil, Mexico, South Korea, Singapore, Taiwan and Hong Kong. The manufacturing growth rate has been very high in the last two decades in these countries. Multinational companies shift production here because of the low costs of labour. The shirt I am currently wearing is from Hong Kong and my trainers from South Korea. Look in any electrical retailer and see where the products are made: 'Japanese' names are even used to hide Singaporean manufacture. Fig. 10.15 shows how the shift to NICs is related to the product life-cycle. The products originate in the Developed World but production is progressively located in the Developing World, giving an **international spatial division of labour.** Is it exploitative or is this pattern desirable?

> **Governments often influence industrial location.**

Some countries, e.g. Malaysia, Philippines and Thailand, are seen as being in the early stages of becoming NICs because industrial investment is growing, e.g. Proton cars in Malaysia are a branch of Mitsubishi. Malaysia is resisting being called an NIC because it feels it to be a derogatory term implying responsibility for deindustrialisation in Europe. Also Malaysia fears it would lose the right to request aid from the Developed World and might have to rely solely on industrial growth for economic development. These moves in industrial production to the Developing World are part of a **global shift** in production.

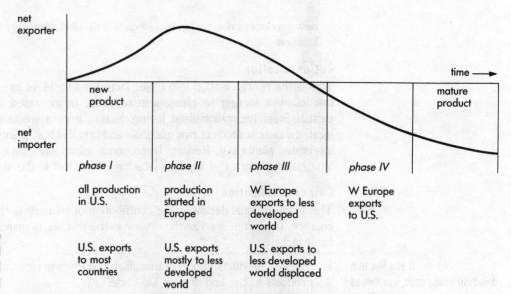

Fig. 10.15 The product life-cycle and its suggested effects on United States and European production and the less developed world

| phase I | phase II | phase III | phase IV |
|---|---|---|---|
| all production in U.S. | production started in Europe | W Europe exports to less developed world | W Europe exports to U.S. |
| U.S. exports to most countries | U.S. exports mostly to less developed world | U.S. exports to less developed world displaced | |

## THE ROLE OF GOVERNMENT IN INDUSTRIAL LOCATION

This can be considered at different levels.

### National government intervention

National government **intervention** in Great Britain began as a result of the Depression years of the 1930s and in the aftermath of the **Barlow Report** (1940) on the *Distribution of the Industrial Population*. At that time Barlow felt that new industry was too focused on the South-East and Midlands, to the detriment of the coalfield industrial areas. In addition, London was being bombed. Therefore government policy to redirect industry to the

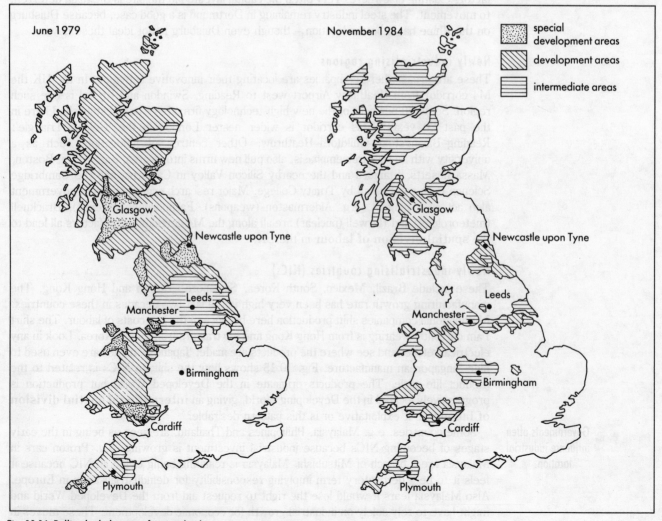

Fig. 10.16 Rolling back the map of regional aid

depressed areas was recommended – a policy which had started in 1934. From 1945 **Industrial Development Certificates** (IDCs) were needed to develop new factories; these were withheld in the South-East and readily granted in the Development Regions. There were three types of **Assisted Area** at the peak of the policy in 1979 in which new firms received all types of assistance to build factories and employ labour: Special Development Areas, Development Areas and Intermediate Areas (Fig. 10.16a). Since 1979 assistance to these areas has been phased out (Fig. 10.16b). The two types of area (Development and Intermediate) were only retained for the purposes of receiving **supra-national** assistance from the **Regional Fund** of the European Community.

### Enterprise Zones

These were established in 1980 to try to improve the economy of inner city areas by creating conditions for industry to develop on derelict sites, although other uses were permitted. Local taxes were remitted for ten years and others reduced, and many of the planning requirements were waived. The Isle of Dogs, London; Speke, Liverpool; and Hartlepool Docks were three of the 27 areas given EZ status (see also chapters 12 and 13).

### Procurement policies

Defence industry expenditure is the largest element of government spending and this can be used to aid industries. The British Navy has placed orders in shipyards in several regions for the same type of vessel, so as to aid the local economies. Increasingly defence expenditure is in high-technology companies and their southern location means that fewer funds go into the northern regions' economies.

### Nationalised industries

Industries owned by the state can be used to assist government industrial policies. Ebbw Vale steelworks was for many years an excellent case of an unprofitable plant heavily dependent on subsidies. Its survival was seen as important for regional/social considerations. Similarly the Ravenscraig and Llanwern plants were built to assist *two* areas rather than concentrating on one area with a larger, more efficient works.

## GOVERNMENT AND INDUSTRIAL DEVELOPMENT IN THE NICs

Industrial development in the NICs can be helped by various measures.

- Governments can demand the processing of primary products before their export. Malaysia has now set up firms to manufacture surgical gloves from rubber.

- Governments may seek to broaden the industrial base by attracting foreign investment in partnership with local capital. This has been the policy of Singapore. When it built its Metro system it was in partnership with foreign companies. Now it has the technology to set up its own company which has made a bid to build the underground in Taipei, Taiwan.

- Governments may seek to build up heavy industry, e.g. iron and steel and cars in Brazil. Private companies had the problem of high-cost production and insufficient domestic markets for some products, e.g. washing machines, because of little affluence in the population as a whole. Government support for such industries was considered vital for them to survive.

- Government support may take the form of high tariff barriers to assist local consumer industries to develop, especially where labour intensive industries will provide many jobs for the local economy.

## CASE STUDY INDUSTRIES

You should be able to discuss the themes of industrial location factors and industrial location changes by choosing a specific industry and by selecting examples from a country, or a number of countries.

### Iron and steel

Iron and steel is **multi-locational** and illustrates how Weberian principles once applied to industry. Themes of transport costs, weight loss, raw materials, inertia, agglomeration and government support and rationalisation can be considered. These themes could be illustrated by changes in the British, West German, French and United States iron and steel industries.

### Brewing

As we noted earlier, brewing is **market-orientated**. The themes which are illustrated by studying its location are: weight gaining, the role of technical developments, linkages *back* to suppliers and *forward* to the retail outlets, transport costs, scale advantages and inertia. Great Britain is the best country to study in this case.

### Car assembly

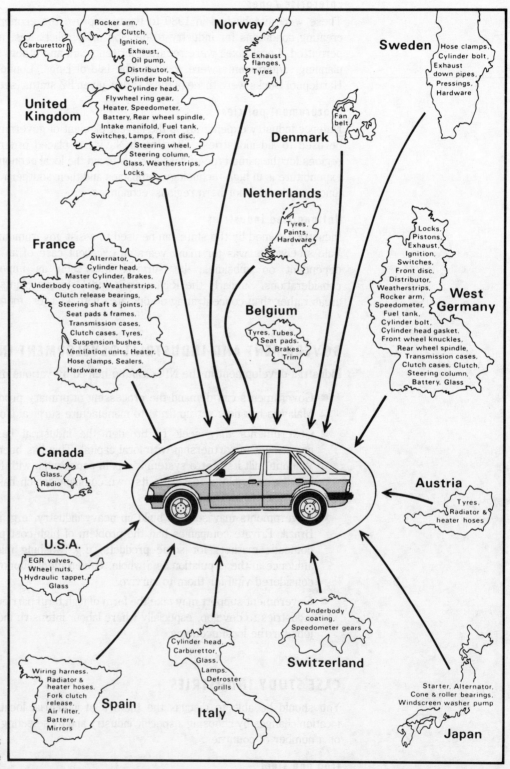

**Fig. 10.17 The component sources for the Ford Escort in Europe**

Car assembly is a **footloose industry** in that its locational preference is less strong. The themes which the motor vehicle assembly industry illustrates are: the unimportance of transport costs and raw materials; the influence of behavioural factors; the influence of markets and trade blocs and of multinational industries; the role of government policy. Once again the British industry provides a good case study; West Germany and the USA would also provide good examples. Motor-vehicle production does illustrate the longest-standing of the transnational corporations and their complex production and locational needs (see Fig. 10.17).

### Electronics

Electronics is a newer and more difficult industry to study. It illustrates many of the more modern factors such as: the role of **multinational companies**; the role of constant **innovation** and new **product cycles**; the part played by **R and D**; the need for a skilled workforce; the low level of transport costs; the concept of the **branch-plant economy**; and the role of government contracts. Great Britain illustrates all of these themes, although you could use other countries.

All of these industries, except brewing, illustrate the **global shift** of manufacturing. By all means select other industries, such as oil refining and chemicals, but make sure that you know *what* is made *where* and *why* it is located there.

### A word of warning

Don't let yourself down by discussing textiles without distinguishing the fibres. Wool, cotton, linen and artificial fibres are spun and woven in different places for different reasons.

## PROJECTS

Industry is not always easy to study because firms are reluctant to discuss sensitive material with the public. However, here are some ideas for an industrial project and the sources of information you might use.

- Industries in a town in the past, using directories as sources of locations.
- The geography of an industrial estate – how have firms changed – newspaper archives might help.
- The impact of industrial closure.
- A government area of assistance – a former Development Area or an Enterprise Zone.

---

**SERVICE INDUSTRIES: THE TERTIARY AND QUATERNARY SECTORS**

Not every syllabus specifically mentions office work, recreation and tourism, retailing, transport and distribution for study. However, in most syllabuses transport is featured as a distinct sector of studies. Retailing normally features in settlement geography and certain aspects of office location also feature as part of settlement geography (e.g. in both the Cambridge and AEB syllabuses). In one form or another, a knowledge of the main trends in the service industries will be a useful background for your geography studies.

## KEY CONCEPTS IN SERVICE INDUSTRIES

### Post-industrial society

This is the other side of the **deindustrialisation** process. It is a description of the modern employment pattern in countries such as Great Britain, France, Australia and the United States of America, where the greater part of the country's economic output and the majority of employment is provided by the tertiary and quaternary sectors.

### Central business district

Office work is normally concentrated in the **central business district** (CBD) of a city and the larger the city, the more disproportionate is that area's share of total office employment. There are three types of office work: (1) the routine, which employs most people, (2) middle management, and (3) the decision-makers and policy-formulators. Each of these groups can have separate locational patterns in a city and its region.

### Producer and consumer services

**Producer services** is the term given to work associated with serving manufacturing, such as accountants and business analysts. **Consumer services** are more concerned with selling the product of manufacturing.

### Office linkage

Offices tend to agglomerate in the centre of the largest cities. London has distinct zones for banking, insurance, commodity trading and stockbroking. Linkage is the name given to the force which pulls closely related functions together, such as banking and finance or estate agents and solicitors. Linkages can be **complementary** – the use of common advertisers; **ancillary** – the provision of joint lunch facilities; **supply** – the use of a common source of office supplies; and **information** – acquiring common knowledge without paying for it.

### Communications technologies

Communications technologies have influenced the nature and locaton of service industries. The manual dexterity required to operate a typewriter was at least partly responsible for office work being seen as a female occupation. The 'office pool', with concentrations of office-based personnel, became a familiar situation. Subsequently, word processors (a term coined by IBM in the 1960s) have had a dramatic effect on the nature of employment and its location. The powerful modern combination of the electronic keyboard, enhanced telephony and telecommunications, allied to the power of the computer, has liberated office work from locational ties to the city centre.

## OFFICES

## UK CASE STUDY: THE DECENTRALISATION OF OFFICES

In the early 1960s the growth in office employment was seen as a major component in the strength of the economy in the South-East. Office employment grew in central London because:

1   Company HQs were located there to gain access to finance;
2   Mergers and takeovers were creating large corporations which wanted head offices in the City;
3   The growing number of transnational/multi-national companies in the service sector wanted a European base in London, e.g. banks;
4   The nationalised industries were all seeking corporate HQs in London;
5   The growing welfare state led to an expansion of the Civil Service.

However, this concentration of office activity in the South-East began to be seen as a 'problem' in the 1960s. The causes for concern were:

a) Congestion caused by extra commuters;
b) Rising land and rental costs in the South-East;
c) Increased labour costs;
d) The effects of property development on the townscape.

The reaction of government was the introduction of the Office Development Permit (ODP) – a similar scheme to the Industrial Development Certificate (IDC). The ODP was designed to push office employment to the regions beyond the South-East. The Location of Offices Bureau (1964–79) was also set up to advise clients on decentralisation and to promote office relocation. In London, some new office centres had been established in development plans for the suburbs. However, only Croydon planned for a major growth of office employment, which was to create 30 000 new and transferred jobs in two decades. The government now began to move Civil Service jobs out of London, e.g. National Savings to Durham, DSS to Newcastle and Motor Taxation to Swansea. Private companies began to move but mainly to towns in the southern half of the country, e.g. IBM to Portsmouth, BP to Harlow, Eagle Star to Cheltenham and NatWest/Sun Life to Bristol.

### The type of jobs moved

■   The **routine work**, which was increasingly involving computerised procedures, could be located anywhere, so long as employees could be persuaded to move. Hence most relocations were to the south, because southern towns were perceived as being more attractive locations by managers in London – the North having a different image! Access was moved to Southend and Barclaycard to Northampton.

The insurance industry, with its routine premium-gathering and payouts, was an ideal routine work operation for decentralisation.

- **Middle management** also moved out – they could use the improving telecommunications networks. These too tended to be located within easy travel of the slimmed-down London HQ for meetings.

- The **decision-makers** largely remained in central locations, because they needed the linkages to money markets and to policy-makers on a face-to-face basis.

More recently, government decisions such as the 'Big Bang' deregulation of the financial markets have brought further pressures to decentralise as new financial institutions have grown and/or come to London (e.g. Japanese banks). The **Enterprise Zone** in London dockland has provided space for some of this decentralisation. Other firms, e.g. Chase Manhattan Bank, have continued the process of relocation, in this case to Bournemouth.

What is the effect of decentralisation on the receiving town and on its workforce – who gains and who loses? Will these moves increase the differences between towns in different parts of the country? These are potential exam questions which you might consider.

## RECREATION AND TOURISM

Here we outline a number of basic concepts used in the study of recreation and tourism.

### Leisure time

This may be occupied by **recreation** activities, undertaken for pleasure, which can be **home-based** or **non-home-based**. It is the latter which is the focus of geographical study, although home-based recreation can have an impact on various activities, e.g. retailing, with the growth of DIY stores. **Tourism** normally involves recreation taking place away from home and involving overnight stays. Tourism is more commercial and generates more wealth because it often involves foreign visits and flows of **invisible earnings** between countries.

### Recreational resources and tourist resources

These terms could include anything which satisfies a demand, ranging from a local park to a country park or from the seaside resorts of the Mediterranean to the open beaches of Pahang, Malaysia, or even the Berlin Wall.

### Primary tourist resources

These are:

- The countryside and coast which people visit for an aesthetic experience (as we saw earlier in the chapter, this is not necessarily the view of farmers!), **National Parks, Areas of Outstanding Natural Beauty** (AONBs), **Nature Reserves, Sites of Special Scientific Interest** and **Heritage Coasts** are all officially designated areas where recreation and tourism are major activities.

- History and heritage sites. Places such as York, Chichester, Florence and Paris all trade on this factor. Much of this aspect depends on the **conservation** of our heritage.

### Secondary tourist resources

These are the facilities which help the visitor to enjoy visiting the primary tourist resources, e.g. hotels, cafés and bars, fun fairs, etc.

### Capacity

Tourist resources all have a **capacity** and this often interests geographers.

- **Physical capacity** This varies with activities – if that capacity is reached it is no longer a pleasure for the participants. Car parks often dictate capacity in the New Forest, and crowd control and safety determine the numbers at a football ground and even the number circulating in Westminster Abbey.

- **Ecological capacity** This is the capability of the natural environment to sustain itself without damage. Such ecological damage as **footpath erosion** has a cost, because someone has to pay for the measures to prevent it. Ecological capacity is increasingly a threat in many ways, e.g. the giant turtles laying eggs at Ratau Abang, Malaysia, each summer are threatened by the tourist hordes and poachers. The

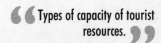
"Types of capacity of tourist resources."

ecological capacity here is very low; will the need for tourist dollars destroy the very environment the tourists come to see?

- **Economic capacity** This is the number of visitors which create the optimum return on a tourist investment.
- **Perceived capacity** This is the individual consumer's own view of the capacity which will be tolerated before the tourist resource has **negative utility** i.e. is no longer enjoyed.
- **Political capacity** It is the task of management to decide on the capacity which is acceptable to all parties interested in a tourist site. Very often this is the task of local government and planning agencies.

## TOURISM IN THE ECONOMY

Tourism acounts for 6 per cent of world trade. Tourism is a major provider of jobs: a quarter of a million people work providing tourist services in London. Tourism has a **multiplier effect** in that the money spreads into the economy to help generate other activities (Fig. 10.18). In some cases the number of people dependent on an activity is very large, e.g. the Cologne Exhibition Centre has only 400 full-time workers yet it has been estimated that up to 40 000 people depend, in full or part, on the Centre for their livelihood.

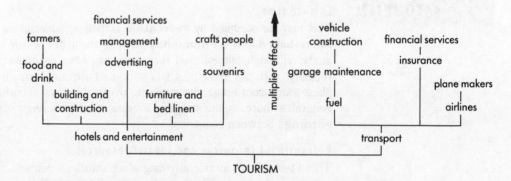

Fig. 10.18 Some aspects of how the multiplier effect works in the tourist industry. The wealth generated indirectly may be two or three times the direct revenue from tourism.

## TOURISM AND DEVELOPMENT

Only 12.5 per cent of tourism is taking place outside of North America and Europe, although tourism is growing fastest in the Developing World. Tourist expenditure does not all flow to the destination country because much of tourism is operated by transnational corporations.    What governs the impact of tourism on an economy (Fig. 10.19)?

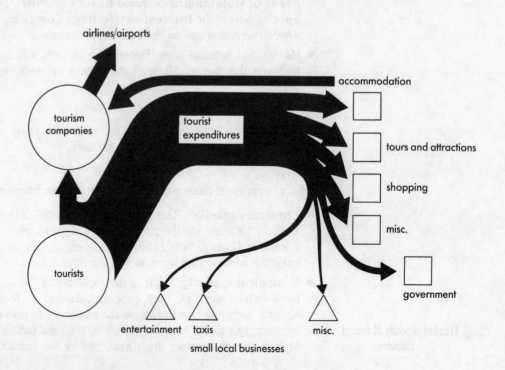

Fig. 10.19 Generalised distribution of tourist industry expenditure in a developing world economy

- It improves the balance of payments and provides wages for the population.
- Local trades and crafts can be stimulated, jobs are created in hotels, bars, taxis and tour services and, unfortunately, sometimes in prostitution.
- Tourism might pull people away from other employment in agriculture, and it can be seasonal. If the employment in an area depends totally on tourism, then there are dangers from:

> *Impact of tourism on the economy.*

1 New resorts creating new supply, e.g. Cancun in Mexico damaged Jamaican tourism.
2 Environmental hazards destroying provision, e.g. 1988 Hurricane Gilbert in Jamaica and 1989 Hurricane Hugo in Antigua and elsewhere, leaving no alternative sources of income.
3 Recession in the Developed World may reduce visitors.
4 Changes in fashion generated by the transnational tourist companies or airlines may have serious local effects: e.g. Quantas moved its flights between Australia and Europe to Bangkok in preference to Singapore, so cutting off a supply of stopovers (i.e. travellers interrupting a journey for additional tourism).

Tourism also has an effect on the host society and its culture which that society may or may not value. It can raise the expectations of the population, e.g. to be better trained. It can reduce the role of the local language in the country because the international language is of greater economic value to the individual. English, German or Japanese will help a Malay more than Bahasi Malay. It can affect religion and moral values, particularly in countries where non-Western religions are indigenous. Musical and theatrical skills may, however, be preserved and possibly adapted, in response to demands for cultural evenings.

For all these reasons, the value of tourism in development should at least be questioned rather than taken for granted.

## PROJECTS

Obviously you are not likely to be able to study the impact of tourism in the Developing World, unless you are lucky enough to spend an extended holiday in the region. Summer holiday fortnights are not usually enough time, given other distractions! However, here are a number of projects you might consider.

- The impact of tourists on footpath erosion.
- The numbers of people visiting and the distance from the car parks of a range of recreational sites.
- Tourist facilities in a resort 1950s and 1990s – how and why have they changed?
- Do visitors to a resort decline with distance from it – does this vary between months and days of week?
- Tourist impact on a landscape – what constitutes overcrowding?
- How perceptions of tourist facilities vary with age and gender.

## TRANSPORT AND COMMUNICATION

These two themes arise in almost all the boards' syllabuses. Transport is the more common theme because it impinges on location. The study of transport may include: network theory; the variety of transport modes and their roles; the choice of locations for transport facilities; theory of port development; and decision-making exercises concerning transport proposals.

*A word of warning*
DO NOT attempt projects on transport provision. The Ministry of Transport cannot get their forecasts correct, even with sophisticated computers, so you will not be able to solve a problem in a few days.

## KEY CONCEPTS AND THEORIES IN TRANSPORT GEOGRAPHY

### Network theory

A transport network comprises (1) **nodes** or **vertices** which are at the meeting point of (2) two or more **edges** or **links**. Nodes are, for example, the railway stations, and edges are the lines. The character of the network depends on the **density** of the network and its **connectivity**, or how well the nodes are connected to each other.

> 66 **Know how to measure connectivity.** 99

- **Measuring connectivity**   Connectivity is measured in three principal ways:
  a) *The cyclomatic number* (u)      $u = e - v + s$
     The cyclomatic number equals the number of edges minus the number of vertices plus the number of **sub-graphs** (subsidiary or unconnected graphs).
  b) *The Beta Index*      $\beta = \dfrac{e}{v}$
     The more edges there are that connect vertices, the greater the connectivity.
  c) *The Alpha Index*      $\propto = \dfrac{u}{(2v - 5)}$
     This compares the maximum number of **circuits** (a path starting and ending at one place) within the graph with the maximum possible for a given number of vertices. It can vary from 0 minimum to 1 maximum.

- **Measuring accessibility** The accessibility of each vertex within a network is measured by:
  a) *The König Number*: This is the maximum number of edges from any one vertex by the shortest path to any other vertex in the network (see the data in the question on Tiree at the end of this section). Lower values show greater centrality.
  b) *The Shimbel Index*: This measures the total number of edges needed to connect any vertex to all others in the network by the shortest path. Again lower values indicate greater accessibility. All of these enable you to compare networks over time e.g. railways in the 1930s and today.

### Transport costs

Transport costs rise with distance but, as Hoover showed, the rises are stepped so that unit costs fall with distance. Costs also vary with the mode of transport. Road transport is cheapest over short distances up to 350 km. Above that rail is cheaper with ship/canal and pipeline proving more expensive still. Air transport is the most costly. However, much depends on the nature of the goods being transported. Fluids are best transported by pipeline or in bulk (supertankers). Bulky goods are best transported by water or rail, e.g. iron ore to the Ruhr. However, more perishable goods and light, small products are more effectively carried on a door-to-door basis by road, or even by air if the product is sufficiently valuable to merit such high-cost transport. It is often easier to provide specialist road transport. Fig. 10.20 shows the modal split of goods transport in the European Community.

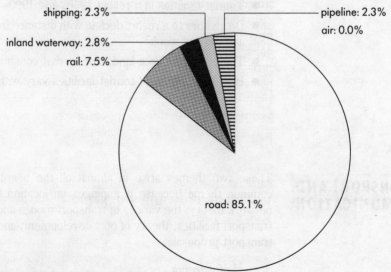

Fig. 10.20 Domestic traffic/goods transport within the EEC (million tonnes), 1981

shipping: 2.3% — inland waterway: 2.8% — rail: 7.5% — road: 85.1% — pipeline: 2.3% — air: 0.0%

### The Taaffe, Morrill, Gould model

This model (Fig. 10.21) was developed to indicate the links between transport and economic development. The model passes through six stages:

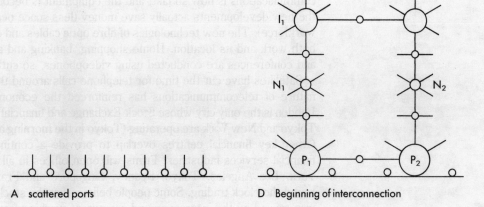

A scattered ports

D Beginning of interconnection

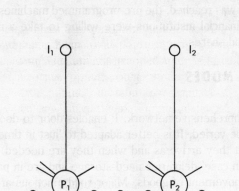

B penetration lines and port concentration

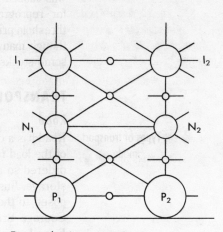

E complete interconnection

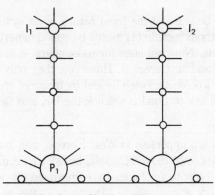

Fig. 10.21 The Taaffe, Morrill and Gould model

C development of feeders

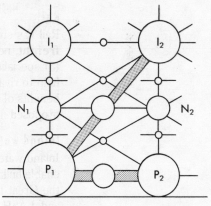

F emergence of high-priority 'main streets'

A   Scattered ports and trading posts along a coast with limited hinterlands.
B   Transport development favours selected ports which expand their hinterland.
C   Nodes develop on the feeders and further advantage gained by the larger ports.
D   The prime routes develop further nodes with their own feeders.
E   The interconnection is complete as all the links made. The nodes are arranged hierarchically.
F   Some routes carry more trade – **main streets** – and develop as trunk routes between the main centres.

You will need a case study of the development of a rail network in the Developing World: Ghana, Liberia and East Africa all make good examples.

### Communications

Communications are the means of transport for ideas and information. Radio, TV, newspapers, telephone and FAX are all elements of communication. These are important for tertiary and quaternary activities, as we saw previously. Because the speed of communications is now so fast, and the equipment is becoming both smaller and cheaper, the new developments actually save money (less space occupied and time savings for the workforce). The new technologies of fibre optic cables and computers have implications for both work and its location. Home shopping, banking and air-travel ticketing are with us, and conferences are conducted using videophones, so cutting travel costs.

Satellites have cut the time for telephone calls around the globe. In addition, the global nature of telecommunications has reinforced the economic dominance of some cities. London is the only city whose Stock Exchange and financial institutions are open when *both* Tokyo and New York are operating (Tokyo in the morning and New York in the afternoon). Other key financial centres overlap to provide a continuous trading potential for the financial services industries. Firms will open offices in all the key cities – London, New York, Los Angeles, Tokyo, Sydney, Singapore and Frankfurt – to take advantage of round-the-clock trading. Some people believe that the stock market crash of October 1987 was caused by the interconnected computers working faster than the people responsible for reprogramming the system, the result being machine-led panic selling! Once a threshold price was reached, the pre-programmed machines gave instructions to sell, even though many financial institutions were willing to take a more cautious view and retain some stocks and shares.

## TRANSPORT MODES

### Road

**" Types of transport mode. "**

Road has a comprehensive network. It enables door-to-door transport and allows the size of the load to be varied. It is better adapted to 'just in time' deliveries (raw materials are ordered so that they arrive as and when they are needed) developed by Japanese to cut storage. 'Just in case' deliveries need storage and are in part a reaction to all the threats posed to the movement of goods. Major road junctions are key locations for factories – Courage Brewery at Reading, hotels and shopping centres (have you got examples?). Even housing that is a short distance from junctions gains more in value. The costs in terms of pollution – lead and noise, vibration, accidents and traffic congestion – are not borne by the traffic.

### Rail

Rail has suffered because of the fixed nature of its system and its high track costs. For **freight**, new methods include (1) merry-go-round mineral trains, (2) containerisation, and (3) specialist trains. New rail lines for passengers, e.g. high-speed trains (French TGV) help to rival road and air transport. However, they only benefit a few places because the benefits of speed would otherwise be lost by frequent stops. The Channel Tunnel and the proposed new rail link to London will link the UK into the expanding European network.

### Inland waterways

Inland waterways are important in West Europe, and in North America. Navigable rivers enable bulk cargoes to be moved cheaply, but slowly: e.g. iron ore to Duisburg, grain from the Great Lakes. New technologies have improved efficiency – push barges, barge trains and LASH (lighters aboard ships). Tributary canals are less profitable than the trunk routes. Waterways attract industry partly for the supply of raw materials but also because of the availability of water for cooling purposes.

### Pipelines

Pipelines for liquids and gases have expanded considerably, e.g. to feed North Sea gas to the UK. There is a high cost in installing the network but the subsequent costs are generally low. Pipelines provide essential links between oil refineries and chemical plants, and aid the distribution of refined as well as crude oil.

### Sea transport

Sea transport is the basis of world trade in primary produce and manufactured goods. Larger vessels for bulk cargoes have altered the location of port facilities. Vessels and port facilities are now increasingly specialised – e.g. roll-on roll-off (RoRo) vessels and container cranes and spider wagons.

### Air freight

Air freight has grown with the increase in lightweight, high-value goods. Out-of-season produce, exotic fruits and flowers and computer parts are examples of such freight. Air passenger transport is also growing rapidly both for business and tourism. A major factor in the growth of freight and passenger carrying has been the technological development of the jumbo jet, with the consequent reduction in carrying costs per kilometre for a given unit of weight.

You should also look at physical, economic and political factors which affect the route taken by various modes.

## AIRPORTS: LOCATION AND EFFECTS

Airports are major users of capital and land, and they need even larger areas beyond for associated facilities. Gatwick Airport employs 40 000 people directly and indirectly. Tour companies using the airport have HQs in Crawley; food is prepared in kitchens on nearby industrial estates; the crews are housed in the surrounding towns, and visiting crews (and delayed passengers!) need hotels. The land area needed for all of these activities is very large.

Road and rail links, car parks, taxi and bus stands all jostle for space. Airports create noise and air pollution, so they must be sited so as to avoid settlements, especially those directly under the flight path. On the other hand, they must be close to the city or area they serve, either in **real (linear) distance** or in **time distance**. The ideal site should be level, as this cuts construction costs – Bombay airport needed hills removed from the flight path.

Airport proposals often result in conflict between government and the local residents. The case for and against airport growth and development has been argued frequently (often violently) in the courts and in the fields: e.g. Narita, Tokyo; Frankfurt, West Germany, and Maplin/Stansted, UK.

## SEAPORT DEVELOPMENT: BIRD'S ANYPORT MODEL (Fig. 10.22)

- Stage A (1)  The primitive port is adjacent to firm ground, yet is where early docks could be excavated: at a bridging point. (2) Quays extend downstream and across the river and (3) become more elaborate.

- Stage B (4)  Wet, tidal docks are built downstream in deeper water with more room onshore and in-river. More warehouses and transit sheds surround the docks.

- Stage C (5)  Linear quays in enclosed, non-tidal docks were built to accommodate larger vessels. The final phase (6) is the building of specialised quays, e.g. roll-on roll-off ferry terminals, container docks and oil terminals, normally further downstream.

There is probably a Stage D which is closing the upstream facilities and the development of alternative uses for redundant quays, e.g. marinas, housing, new employment nodes. If you live in London and know the docks, see if you know which docks fit into this model. Try it on the evolution of other ports and note the exceptions to the model.

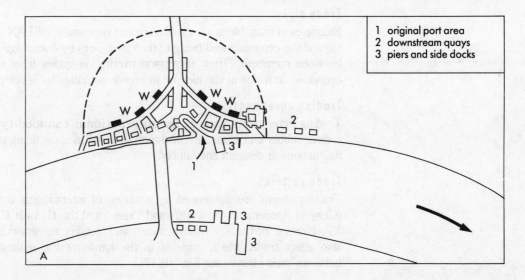

1 original port area
2 downstream quays
3 piers and side docks

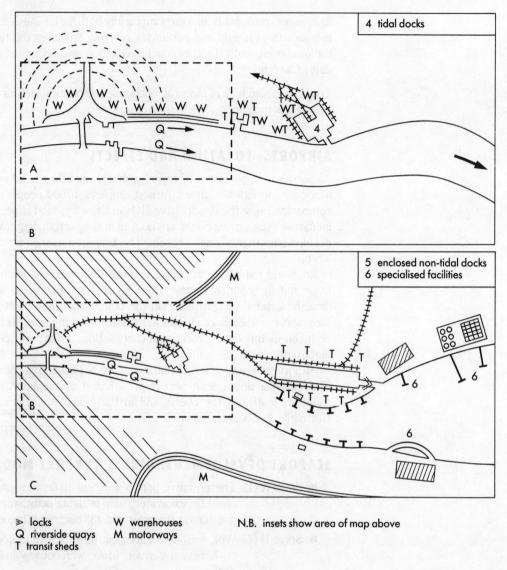

Fig. 10.22 Bird's Anyport model

≫ locks    W warehouses     N.B. insets show area of map above
Q riverside quays    M motorways
T transit sheds

## TRADE PATTERNS

## SOME KEY IDEAS

### Trade

**Visible trade** is the flow of commodities from producers to consumers; **invisible trade** is the flow of services from originators to recipients. Trade is dominated by the Developed World and its need for raw materials, energy and food. It is greater in value between developed countries.

### Trade blocs

Examples of trade blocs are the European Community, ASEAN and Caricom. These aim to expand the economic well-being of their members by encouraging the **free trade** of goods between members. They also place **tariffs** on goods from outside, either to protect economic activities in the bloc or to enable activities to develop.

### Trading agreements

Trading agreements such as the **International commodity agreements** for cocoa, coffee, rubber and sugar, are designed to assist in maintaining price stability in the face of fluctuations in demand and supply.

### Trade patterns

Trade patterns are influenced by a series of international institutions, such as GATT (General Agreements on Tariffs and Trade) and the **Brandt Commission** (see chapter 13). Finance for trade, exchange rates, and stability and growth of each national economy also affect trade. Much trade is in the hands of the multinational companies trading between their plants (see Fig. 10.17).

# EXAMINATION QUESTIONS

## QUESTIONS ON AGRICULTURE

### Question 1

Using specific examples, illustrate the factors which will distort the pattern of agricultural land use developed by J.H. von Thünen.

### Question 2

a) Define 'intensive agriculture'. (3)

b) According to von Thünen, why is it that the intensity of agriculture increases nearer the city? (6)

c) Geographers such as Sinclair have suggested that this trend in agricultural intensity does not always apply. Give the reasons which support this suggestion. (6)

(AEB, June 1988)

### Question 3

With reference to a number of examples at different geographical scales, assess the usefulness of elementary agricultural location theory in explaining real-world patterns of agricultural and rural land use.

(O and C)

### Question 4

Quoting examples from more than one country in Africa, discuss the extent to which variations in agricultural production from year to year have been influenced by:

a) rainfall variability;
b) pests and disease;
c) changes in market demand;
d) political unrest.

(JMB 'C', Alternative D)

### Question 5 (A data response question)

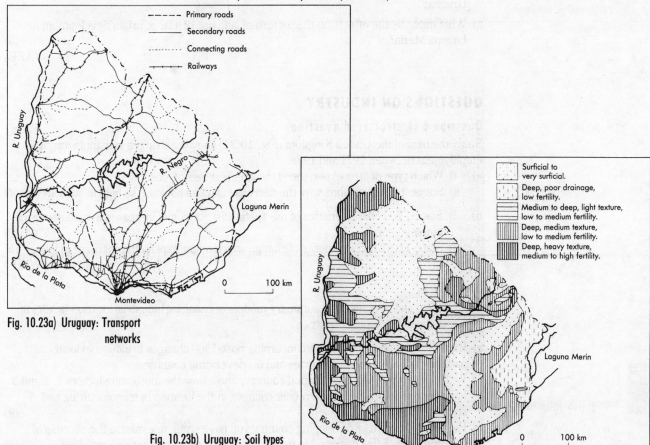

Fig. 10.23a) Uruguay: Transport networks

Fig. 10.23b) Uruguay: Soil types

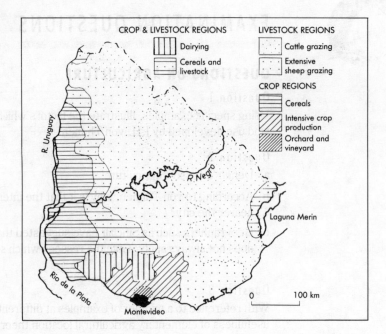

**Fig. 10.23c) Uruguay: Agricultural and land-use regions**

CROP & LIVESTOCK REGIONS
- Dairying
- Cereals and livestock

LIVESTOCK REGIONS
- Cattle grazing
- Extensive sheep grazing

CROP REGIONS
- Cereals
- Intensive crop production
- Orchard and vineyard

0        100 km

Study Figs 10.23 a), b) and c) and the following quotation from P. Ward English's text *A Question of Place.*

'Uruguay is a nation of a mild climate, a low rolling terrain, and rich grasslands. Uruguay's population of 3 million is racially European and its economy deeply anchored in animal husbandry. Half of its population is in the primate city of Montevideo.'

a) Draw a diagram to show how the model developed by von Thünen can be modified to fit the case of Uruguay. (5)

b) Describe and attempt to account for the similarities and differences between your modification of von Thünen's model and the pattern of agricultural land use regions in Uruguay. (15)

a) What might be the effects on the pattern of land use of a large urban development at Laguna Merin? (5)

(AEB)

## QUESTIONS ON INDUSTRY

### Question 6 (A structured question)
Study the map of the United Kingdom (Fig. 10.24) showing changes in manufacturing employment between 1971 and 1978.

a)  i) Which type of area experienced the greatest loss of jobs? (2)
    ii) Suggest TWO reasons why the decrease in jobs should be so high in such areas. (4)

b)  i) State TWO characteristics of the spatial distribution of increases in manufacturing employment. (2)
    ii) State THREE components of change which might explain this increase in jobs. (6)

(WJEC)

### Question 7
Fig. 10.25 a) and b) shows two summary models of changing industrial location factors in developed and developing countries.

a) Explain what the models show concerning post-1960 changes in industrial location patterns in (i) developed countries and (ii) developing countries. (8)

b) With reference to ONE developed country, show how the motivating factors 1, 2 and 3 have been responsible for the recent changes in the location of manufacturing and service industries. (9)

c) For any developed *or* developing country you have studied, evaluate the success of government policies designed to develop new industries and service activities. (8)

(London 16–19, AS Sample)

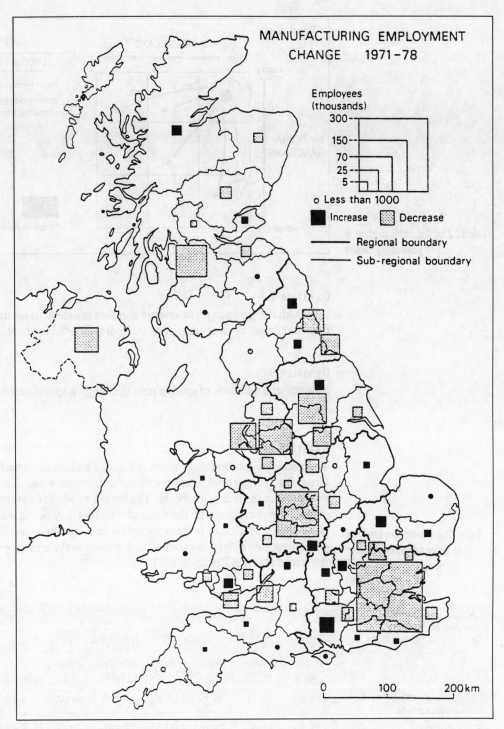

Fig. 10.24

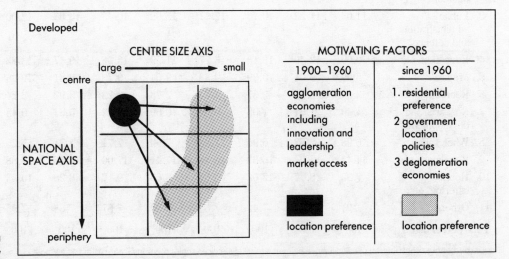

Fig. 10.25a) Industrial location in a developed country

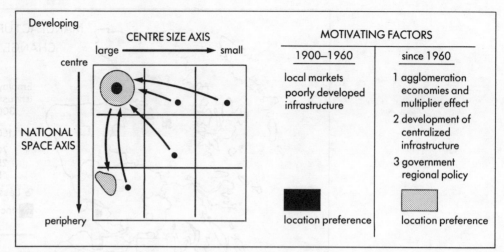

Fig. 10.25b) Industrial location in a developing country

## Question 8

'Although the part played by energy supplies in industrial location decisions at the present time should not be over-stressed, some of their effects are still strongly evident.' Discuss.

(O and C)

## Question 9

Examine the influence of government in changing spatial patterns of industrial location and growth.

(O and C)

## Question 10

The relative importance of various industries and services within individual regions of Great Britain may be similar to or different from the relative importance of the same industries in the national context. The extent to which each region's mixture of industries or services compares with the national situation may be measured by the index of dissimilarity which may be interpreted on the diagram below. Table 10.2 shows the percentages employed in various categories of service employment for each region and for the whole of Great Britain in 1981.

Table 10.2 Percentages employed in major categories of service occupation by region and nationality, 1981

| Major category of service occupation | Region | | | | | | | | | | Great Britain[1] |
|---|---|---|---|---|---|---|---|---|---|---|---|
| | South-East | East Anglia | South-West | West Midlands | East Midlands | Yorks and Humber-side | North-West | North | Wales | Scotland | |
| 1. Transport and communication | 12.07 | 10.95 | 8.79 | 8.40 | 9.99 | 10.28 | 10.66 | 9.70 | 9.55 | 10.22 | 10.68 |
| 2. Insurance/banking | 17.03 | 11.20 | 11.64 | 11.76 | 10.37 | 10.46 | 11.66 | 9.49 | 8.85 | 10.58 | 13.21 |
| 3. Public administration and defence | 11.03 | 11.22 | 10.70 | 12.51 | 13.99 | 10.43 | 13.01 | 12.00 | 17.76 | 12.20 | 11.92 |
| 4. Education | 10.00 | 12.49 | 12.18 | 13.77 | 11.59 | 13.96 | 10.47 | 13.22 | 12.41 | 11.03 | 11.35 |
| 5. Medical | 8.59 | 9.04 | 10.29 | 9.33 | 9.64 | 9.78 | 9.81 | 10.06 | 11.52 | 11.81 | 9.58 |
| 6. Recreation | 3.30 | 2.55 | 2.57 | 2.74 | 2.78 | 3.22 | 3.92 | 3.74 | 3.17 | 3.56 | 3.25 |
| 7. Research and development | 1.30 | 1.61 | 1.05 | 0.39 | 0.69 | 0.42 | 0.84 | 0.43 | 0.27 | 0.67 | 0.92 |
| 8. Wholesaling | 6.98 | 7.83 | 6.96 | 7.92 | 7.90 | 7.71 | 7.31 | 5.20 | 5.73 | 5.42 | 6.94 |
| 9. Retailing | 14.41 | 17.24 | 16.32 | 16.48 | 17.41 | 17.00 | 16.16 | 17.88 | 14.14 | 15.63 | 15.64 |
| 10. Hotels, catering, etc. | 7.38 | 8.47 | 12.08 | 9.07 | 8.65 | 9.17 | 8.28 | 10.33 | 9.37 | 10.16 | 8.71 |
| 11. Others | 7.91 | 7.40 | 7.42 | 7.63 | 6.99 | 7.57 | 7.88 | 7.95 | 7.23 | 8.72 | 7.80 |
| Total percentage | 100 | 100 | 100 | 100 | 100 | 100 | 100 | 100 | 100 | 100 | 100 |

[1] i.e. percentage contribution of each service category to total service employment in Great Britain as a whole.

a) Complete Table 10.3 by calculating the indices of dissimilarity for the following five regions: West Midlands, North-West, North, Wales and Scotland. The formula you require is given below Table 10.3.

Table 10.3

| Region | South-East | East Anglia | South-West | West Midlands | East Midlands | Yorks and Humber-side | North-West | North | Wales | Scotland |
|---|---|---|---|---|---|---|---|---|---|---|
| Index of dissimilarity for manufacturing | 14.01 | 19.46 | 17.28 | 30.88 | 8.10 | 19.82 | 9.76 | 16.12 | 12.77 | 15.16 |
| Index of dissimilarity for services | 5.97 | 4.59 | 5.74 | | 5.10 | 5.40 | | | | |

The index of dissimilarity should be calculated as follows: name the two sets of percentages to be compared $p$ and $q$ where $p$ represents the set of percentages for a given region and $q$ represents the set of percentages for Great Britain. Select only the pairs where $p$ is greater than $q$ and apply the formula – index of dissimilarity = $\Sigma\,(p-q)$, where $\Sigma$ = sum of.

Show your full workings in your answer book.

0                    Index of dissimilarity                    100

Region identical to national situation

Region as different as possible from national situation

(5)

b) Using Table 10.3 draw two choropleth maps on the base maps provided (Figs. 10.26 a and b), one to show the regional variations in the index of dissimilarity for manufacturing employment and the second to show the regional variations in the index of dissimilarity for service employment. You should choose a suitable number of categories to show the variations for each map separately. Ensure that each map has a title and a key. (8)

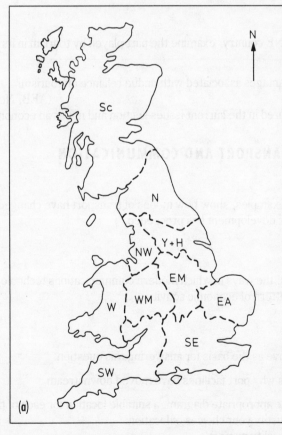 

Fig. 10.26

c) Using the maps you have drawn and the completed Table 10.3, comment on the differences and/or similarities between the values shown for manufacturing and the values shown for services. (7)

(JMB 'B' and 'C' Practical, June 1988)

## QUESTIONS ON OFFICES

### Question 11

Explain how and why the location of offices and community services, such as hospitals and schools, changes over time both within and between settlements.

(O and C)

### Question 12

a) Describe the office location pattern of a typical city in the Developed World. (*10*)

b) Account for the pattern which you have described. (*15*)

(AEB, June 1989)

**NB** This question was set in the settlement section and has a settlement bias.

### Question 13

The location of tertiary activities is determined by accessibility. Discuss.

(London 210)

**NB** This question is broader than offices and includes tourism and retailing.

## QUESTIONS ON TOURISM

### Question 14

a) Describe the attractions of Southern Asia for the international tourist. (*9*)

b) With reference to at least TWO countries from Southern Asia discuss (1) the benefits of tourism to their economies and (2) the environmental, economic and social costs that may occur as a result of tourist development. (*16*)

(JMB 'C')

**NB** Southern Asia is south of Turkey, USSR and China. This is a question from a regional syllabus, but you could get a similar question which is open-ended, i.e. leaves you to choose the countries.

### Question 15

a) With reference to ONE country, examine the part played by tourism in its economic development. (*15*)

b) What are the disadvantages associated with undue reliance on tourism? (*10*)

(AEB, November 1987)

**NB** This question featured in the current issues section and not as an economic question.

## QUESTIONS ON TRANSPORT AND COMMUNICATION

### Question 16

Using local or regional examples, show how modes of transport have changed in importance as economic development has proceeded. (*25*)

(Oxford)

### Question 17

Discuss, with examples, the ways in which modern communications technology is affecting the global pattern of economic activity. (*25*)

(Oxford)

### Question 18

Turn to Figure 10.2 above as the basis for answering this question.

a) List THREE reasons why port facilities have moved downstream. (*3*)

b) Mark and label on the appropriate diagram, a suitable location for each of the following facilities. Justify your choices of location:
  i) a modern, high-quality marina; (*3*)
  ii) a freeport facility; (*3*)
  iii) a roll-on/roll-off ferry terminal; (*3*)
  iv) a container terminal/port. (*3*)

(AEB, 1987)

## STUDENT'S ANSWER WITH EXAMINER'S COMMENTS

### Question 1

> **Good introduction.**

In 1820s Johann von Thünen a German agriculturalist devised a model that showed the pattern of agricultural land-use in a given area. His chief main variable was the distance from the market. He was mainly interested in the economic rent of a particular crop, i.e. where a particular type of farming, whether arable or pastoral, gives greatest returns in relation to distance and transport costs to the market.

> **Good outline of assumptions. It is worth briefly outlining the model.**

In this model, von Thünen made several assumptions so to make the model work. The first one was that all farmers are economic men, i.e. that they are optimisers not satisficers. The second, that transport costs are proportional to distance. The third, there is only one market centre in a given area. And the fourth, that the chosen area is a isotropic plain, i.e. soil fertility, climate, topography of land are all constant.

> **Good point.**

When his model was first discovered in the 1820s many of the factors that were important then, are no longer so, i.e. as there was no refrigeration, all perishable goods had to be located near to the market so that the goods would not rot or go off.

> **Give examples.**

Transport costs are no longer proportional to distance, because in von Thünen's time there was only horse and cart to deliver the goods, nowadays there are many different methods of getting goods to the market.

> **A diagram could have been used here.**

The factor of aspect and topography i.e. not an isotropic plain can greatly effect the pattern of agricultural land-use. If we look at Sydney, Australia as an example we can see that the most intensive forms of agriculture are found nearer to the city, as these are sometimes the more perishable goods and the most bulky to transport, i.e. high transport costs. If they were located further away from the market where the land is less urbanised, sparce and cheap, the extensive types of farming are found, i.e. sheep grazing. From this diagram, it does not tell us whether topography has effected where the different types of agriculture are sited. From a more complex diagram of von Thünen's model, we can see the changes.

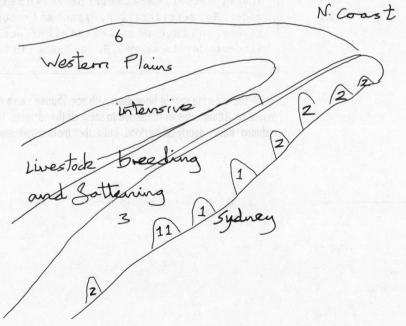

> **Good to see this diagram.**

> **Rather vague.**

From this diagram, we can see that it just isn't milk farming near to the market but also beef agriculture, in the areas where there is no other use for it. Sydney is covered with hills and unfertile/fertile land in a vast number of different places and this can determine what the agricultural land-use will be, you would think that cream and beef farming would be found on the outskirts of Sydney, but from diagram 1 you can see that there is a larger section containing beef agriculture but there is no areas for sheep and crops. This is because of the Western plains which have not got the right conditions for the crops as the gradient of the land is too high, therefore machinery i.e. tractors would find it very dangerous to work here.

Having a navigable river flowing through an area can distort agricultural land-use patterns. As in Addis Ababa, Ethiopia, the vegetation was grown near to the city, but because of the climate, it tended to be located along the river valleys.

Having a second market can mean that instead of having 2 sets of 6 concentric circles, for example

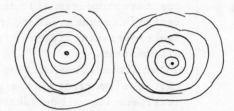

you may get overlapping as there is not enough land for individual land use. This factors would distort the pattern of land-use developed by J.H. von Thünen.

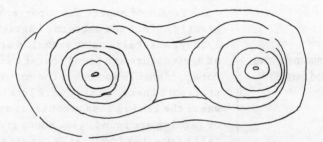

From these factors we can stay that there are many criticisms to von Thünen's model, but it still proves to be the most useful geographical model. So, refrigeration, aspect and topography, fertility of land, rivers, and a second market can all effect the pattern of agricultural land-use developed by J.H. von Thünen in the 1820s.

CHAPTER

11

# SETTLEMENT

## GETTING STARTED

Settlement geography is one of the most popular sections of most syllabuses so far as one can judge. In one board over 40 per cent of the projects attempted are based on this section of the syllabus. Most of you live in urban settlements and therefore it is reasonable for the examiners to assume that you have kept your eyes open while you are out and about and that you might be able to use your home city, town or village to provide examples. Therefore, try to apply as many of the key concepts, theories and definitions as you can to your home settlement. In this way your 'example bank' is doubled.

Some syllabuses (such as NISEC, Oxford, and Oxford and Cambridge) focus on *particular aspects* of settlements within their optional structure. The functions, issues, conflicts and plans for our settlements are a very strong aspect of much of the London 16–19 syllabus, and feature in their decision-making exercises and optional modules. Settlements often form part, if not all, of questions which are based on Ordnance Survey maps and, for that reason, the sample questions include one which contains a 1:50 000 map and two oblique aerial views. The maps of urban areas which might appear in your examination can be large-scale (1:10 000) or perhaps include some of the new city maps.

Settlement geography includes both urban and rural settlement and these can form separate, distinct, questions although there is scope for overlap. Rural studies normally comprise about 20 per cent of your studies in settlement grography.

**KEY CONCEPTS IN SETTLEMENT GEOGRAPHY**

**MODELS OF URBAN STRUCTURE**

**LAND VALUES IN CITIES**

**THE SIZE AND SPACING OF CITIES**

**PLANNING**

**THE PATTERN OF FUNCTIONS IN A CITY**

**URBAN POLICY ISSUES**

**RURAL SETTLEMENT**

**COMMUTER VILLAGES**

# ESSENTIAL PRINCIPLES

## URBAN GEOGRAPHY

This focuses upon cities and towns and is the description, analysis and explanation of both the pattern of urban places and the arrangement of functions within those places. Therefore it not only focuses upon the *processes* which change urban places but also on the *people* behind the processes, particularly the planners. The processes of urban change affect both the economic and social geography of urban places.

## RURAL SETTLEMENT

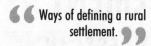
> Ways of defining a rural settlement.

This normally refers to places which are smaller than towns – e.g. villages, hamlets and isolated farms – and which are agricultural settlements. The study of rural settlement also involves the description and analysis of the **pattern** of places and the **form** or **shape** of these places. However, because most places in the countries of the Developed World now house people who work in urban areas or whose life-style is urban, rural settlement studies now tend to examine the relationship between urban and rural settlements and the internal patterns of housing within villages. Today many see the rural settlement as being defined by size, on which there is no agreement, or on the population's perception of the nature of the settlement. However, many people, such as suburban dwellers, the property industry and farmers, have differing ideas of what exactly it is that constitutes a village.

## URBANISATION

This is the process of becoming an urban society. It is a process which has had different causes and consequences at different periods of time and in different places. The process involves:

1 A rapid growth of the town or city's population, e.g. Cardiff grew from 1870 people in 1801 to 164 333 by 1870;
2 An equally rapid expansion of the urban area; and
3 An expansion of the population in secondary occupations (see chapter 10).

Urbanisation involves **rural to urban migration**, combined with a rapid natural increase of the population (see chapter 3).

Urbanisation is also a **social process** because it changes a rural society into a more stratified one, where people are wage-earners and living separately from the owners of the new factories. All this is best applied to the nineteenth century in Europe and to the late nineteenth and early twentieth century in the USA.

Urbanisation in the **Developing World** is a later phenomenon. The causes of the rural push might be similar to those experienced in Europe (do you know them?), and the pull of life in big cities might be identical, but the *speed* and *scale* of the process is different. It often results in 'urbanisation without industrialisation' (see chapter 13) because it has *not* been based entirely on the growth of secondary employment but on hopes, myths and dreams. It often results in **underemployment** – too many people doing too few jobs – in rural areas merely being replaced by **unemployment** – involuntary lack of work – in the towns.

Urbanisation in the socialist world has often been *planned* around new settlements. However, in China and Kampuchea there have been attempts to stem the growth of cities because they have been considered to be capitalist, rather than communist, settlements.

> Be familiar with the cycle of urban development.

### Cycle of urban development
There is a sequence or **cycle** of urban development in population terms (Fig. 11.1).

### Suburbanisation
This is the decentralisation of people and their associated employment, normally from the central districts of a city to the margins.

### Counter-urbanisation
This is the movement of people and employment to the small towns and villages outside the city. It was first noted by Berry in the USA. The movement can be from the city and its

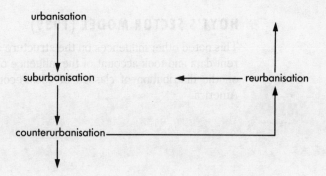

Fig. 11.1 Cycle of urban development

suburbs or directly from other city fringes. It is normally associated with the growth of tertiary and quaternary employment (see chapter 10).

### Re-urbanisation

This is the movement of people and jobs *back* to areas of cities which had been either (1) abandoned during the process of suburbanisation and counter-urbanisation or (2) left derelict as a result of **deindustrialisation** (see chapter 10). The returning numbers and jobs are different from the period of urbanisation – e.g. financial services at Canary Wharf replacing dock work, and the young, affluent, professionals replacing the dock workers in Wapping.

## MODELS OF URBAN STRUCTURE

## THE BURGESS MODEL (1925)

The **concentric zone model** was based on ideas developed by the Chicago School of Urban Sociology and is the earliest structure model. The model (Fig. 11.2) is based on a particular period in North American history and on ecological principles applied to people. So **invasion** and **succession** of people and activities lead to new dominant groups and activities in areas, all fuelled by (1) growth of the city economy in the centre and **zone in transition** and (2) the migration of people to the city. Burgess is a starting point, and his model is not the most useful available today. Any project that attempts to see many similarities between the Burgess model and a modern town or city is unlikely to be well founded. The following models have developed this early work.

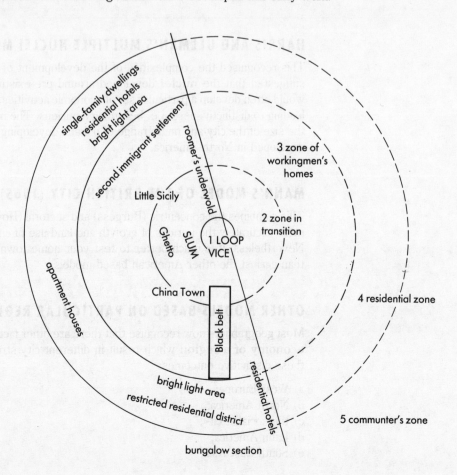

Fig. 11.2 The Burgess model applied to 1920 Chicago

## HOYT'S SECTOR MODEL (1939)

This noted other influences on the structure of the city (Fig. 11.3). His study was based on rent data and took account of the influence of **transport routes** on economic activity and on the distribution of classes of socio-economic groups. It, too, was based on North America.

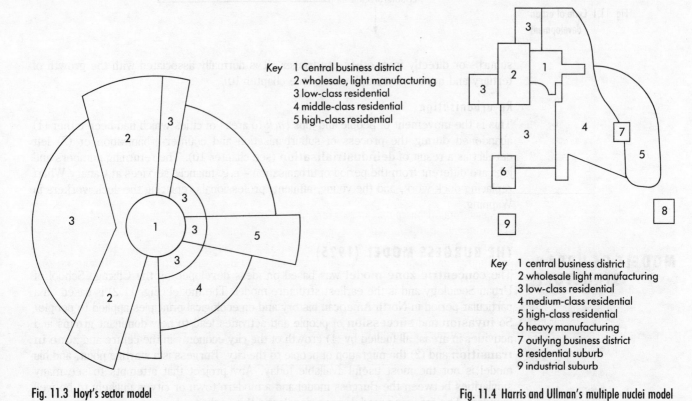

Key  1 Central business district
2 wholesale, light manufacturing
3 low-class residential
4 middle-class residential
5 high-class residential

Key  1 central business district
2 wholesale light manufacturing
3 low-class residential
4 medium-class residential
5 high-class residential
6 heavy manufacturing
7 outlying business district
8 residential suburb
9 industrial suburb

Fig. 11.3 Hoyt's sector model

Fig. 11.4 Harris and Ullman's multiple nuclei model

## HARRIS AND ULLMAN'S MULTIPLE NUCLEI MODEL (1945)

This recognised the complexities of the development of modern cities (Fig. 11.4). They suggested that the **nuclei** developed around pre-existing foci (villages). Each of these would often develop a particular set of economic activities or a unique social attractiveness, leading to distinctiveness, i.e. becoming a nucleus. The number of such nuclei depends on the size of the city and on its range of old and developing functions. The theory was again developed in North America.

## MANN'S MODEL OF THE BRITISH CITY (1965)

This combines the concentric (Burgess) and sectoral (Hoyt) models (Fig. 11.5). It reflects more accurately the periods of growth and land use of cities, although it is already dated. Nevertheless, it is much better to test your home town or city against the Mann model than against the other American based models.

## OTHER MODELS BASED ON PARTICULAR REGIONS OF THE WORLD

Most geographers now recognise that there are other factors **unique to the culture and economy of a region** which result in different city structures. Fig. 11.6 shows five of these distinctive outcomes:

a) West Europe;
b) North America;
c) the socialist city;
d) Latin America;
e) South-East Asia.

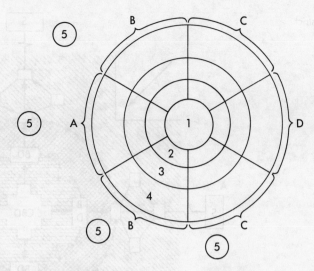

A middle-class sector
B lower-middle-class sector
C working-class sector (and main municipal housing areas)
D industry and lowest working-class areas

1 city centre
2 transitional zone
3 zone of small terraced houses in sectors C and D; larger by-law houses in sector B; large old houses in sector A
4 Post-1918 residential areas, with post-1945 development mainly on the periphery
5 commuting-distance 'village'

Fig. 11.5 Mann's model of the British city

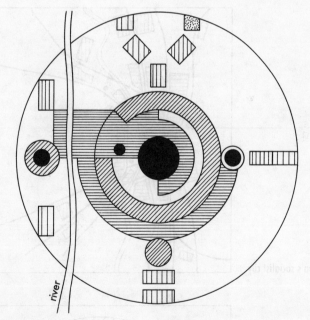

URBAN ZONE AGGLOMERATION AREA

central area and business district

mixed residential and industrial

central residential and recreational

URBAN AREA

built-up suburbs

urban peripheral development

self-standing urban sub-centre

TRANSPORT AND INDUSTRIAL ZONE

urban peripheral development

SUBURBAN ZONE

existing or developing middle or lower order centre

recreational settlement

strongly suburbanised village

other urbanised village

Fig. 11.6a) Nellner's model of the West European city

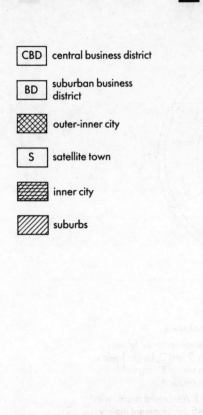

CBD    central business district

BD    suburban business district

⌗    outer-inner city

S    satellite town

▨    inner city

▧    suburbs

Fig. 11.6b) Wreford-Watson's North American city

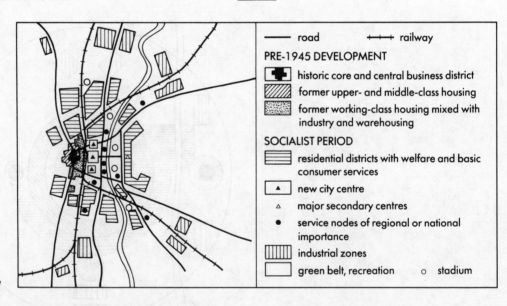

Fig. 11.6c) Hamilton's socialist city

road            railway

**PRE-1945 DEVELOPMENT**

✚    historic core and central business district

▨    former upper- and middle-class housing

▨    former working-class housing mixed with industry and warehousing

**SOCIALIST PERIOD**

☰    residential districts with welfare and basic consumer services

▲    new city centre

△    major secondary centres

●    service nodes of regional or national importance

⦀    industrial zones

☐    green belt, recreation        ○    stadium

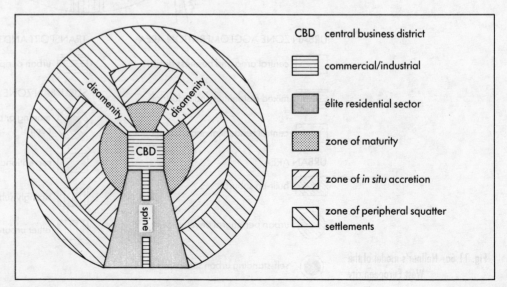

Fig. 11.6d) Cities in Latin America

CBD    central business district

☰    commercial/industrial

▨    élite residential sector

▨    zone of maturity

▨    zone of in situ accretion

▨    zone of peripheral squatter settlements

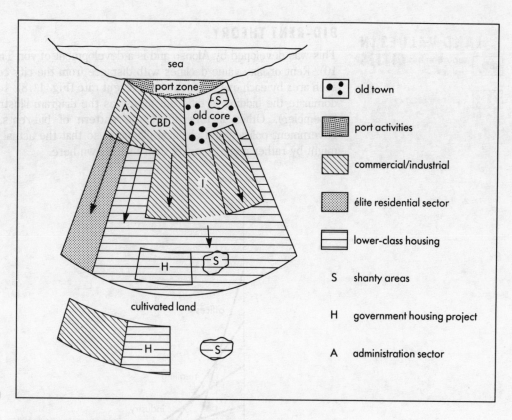

| | |
|---|---|
| ● ● ● | old town |
| (shaded) | port activities |
| (hatched) | commercial/industrial |
| (dotted) | élite residential sector |
| (lined) | lower-class housing |
| S | shanty areas |
| H | government housing project |
| A | administration sector |

**Fig. 11.6e)  Cities in South-East Asia**

## MURDIE'S MODEL OF SOCIAL AREAS (Fig. 11.7)

This attempts to link the three major elements of social area analysis, namely the ethnic, family, and socio-economic dimensions of **social space**, to **physical space**.

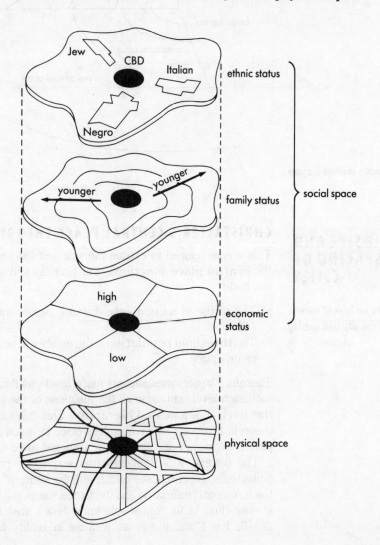

**Fig. 11.7  Murdie's model of social space: Toronto**

## LAND VALUES IN CITIES

### BID-RENT THEORY

This was developed by Alonso and is a development of von Thünen's model (see chapter 10). Rent or land value declines with distance from the city centre. The value placed on each area by each use declines at a different rate (Fig. 11.8), so permitting certain uses to dominate the bidding in different areas, as the diagram illustrates (which model does it resemble?). Other factors affect the pattern of bid-rents, notably transport, local government policies and physical planning, so that the actual pattern of bid-rent curves might by rather more complex than that shown here.

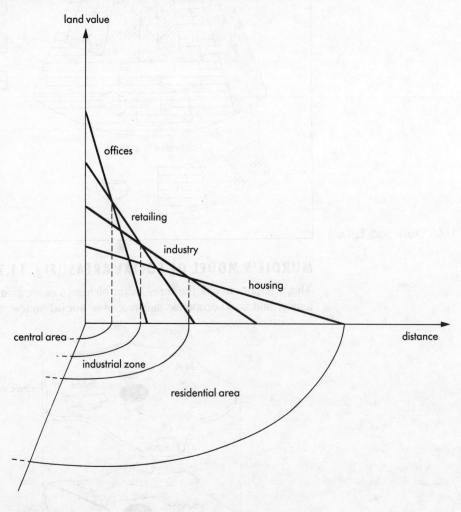

Fig. 11.8 Alonso's bid-rent diagram

## THE SIZE AND SPACING OF CITIES

> " Here we look at various models of the size and spacing of cities. "

### CHRISTALLER'S CENTRAL PLACE THEORY (1933)

This was developed to explain the size and spacing of settlements in an area by studying the **central place functions**, i.e. retailing and services, in settlements. Each function has both:

a) The **range of a good** – the distance people will travel to obtain that good or service; and

b) The **threshold population** – the number of people needed to support any function in a central place.

Therefore larger cities support more functions than small towns and a **hierarchy** exists, with each level containing all the functions of the lower level *plus* the ones dependent on that level – it is a **nested hierarchy**. Christaller also assumed an **isotropic surface** – a theoretical concept of a featureless place in which physical conditions, population density, purchasing power and access are assumed to be uniform.

The pattern of settlements was arrived at geometrically as a series of hexagonal hinterlands, superimposed on a triangular lattice of settlements – the relationship between the hexagonal hinterlands and the lattice varies according to the principle of organisation or k value (Fig. 11.9). You should know how Christaller's theory is modified both in theory (k = 3, k = 4 and k = 7) as well as in reality by, for example, political boundaries, resources, physical geography, human behaviour and choice, and planning, and have examples of each modification. Do you know where the theory has been tested and proved?

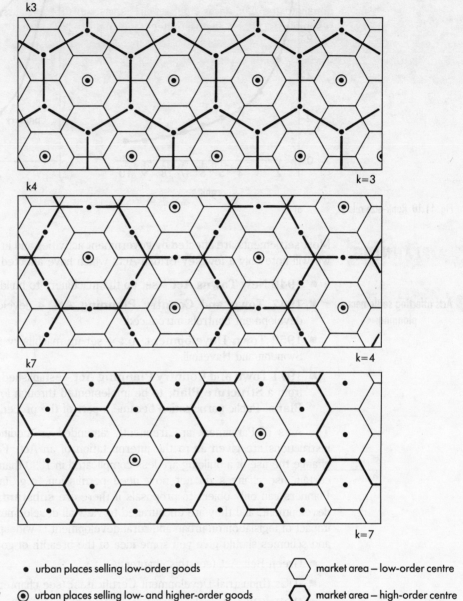

Fig. 11.9 Central place theory

- ● urban places selling low-order goods
- ◉ urban places selling low- and higher-order goods
- ⬡ market area – low-order centre
- ⬡ market area – high-order centre

Lösch (1954) developed the theory by taking a greater range of k values and superimposing them on one another and rotating them to produce **city rich** and **city poor** sectors. He also noted that *not all* central places need to have all of the functions of lower order settlements. This is a better mirror of reality.

## THE RANK–SIZE RULE OF ZIPF (1949)

This states that if the settlements of a country are ranked, then there is a size relationship to the largest city of the form $P_n = \dfrac{P_i}{n}$, such that $P_n$, the population of a city ranked nth, is equal to $P_i$, the population of the *largest city*, divided by its rank, n. This describes the size of cities, when plotted on graph paper, as a 'J-shaped curve' (Fig. 11.10a), but it does not *explain* the pattern. On Log/Log graph paper a perfect rank–size distribution will assume a straight line (Fig. 11.10b). In reality the relationship is rarely perfect and can be **primate**, as in many Developing World countries (see also chapter 13), or **binary**, where two cities dominate, so making the rank–size pattern stepped. A stepped rank–size hierarchy is more akin to Christaller. This theory says nothing about the spacing of settlements.

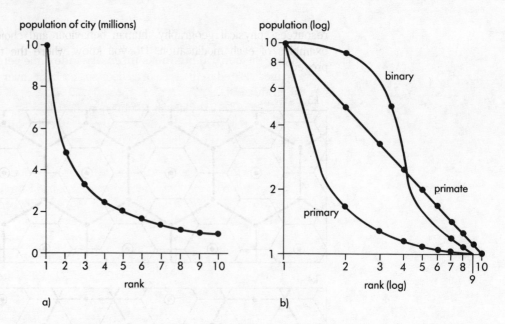

**Fig. 11.10 Rank–size rule**   a)                                    b)

**PLANNING**

Most settlements are affected by government activities and in particular by planning. Here we will note only a few key British Acts which have affected settlement planning.

66 **Acts affecting settlement planning.** 99

- **1946 New Towns Act** – set up the machinery to build new towns, e.g. Crawley.
- **1947 Town and Country Planning Act** – development plans required and development controls introduced.
- **1952 Town Development Act** – set up machinery for **expanded towns**, e.g. Swindon and Haverhill.
- **1971 Town and Country Planning Act** – established a hierarchy of plans, ranging from a **Structure Plan**, to be implemented through lower-tier **Local** or **District Plans**. Public participation became a part of the procedures.

The Acts of Parliament are frequently amended or modified by government or new instructions are given as to the interpretation of an Act. For instance, permissions to change the use of a building are less complicated in 1990 than they were in 1975 because certain use changes do not now need permission, e.g. factory to retail warehouse. Planners can only object to proposals if there are **substantial grounds** to oppose the development, and they are encouraged to view all development proposals positively. The impact of legislation on urban and rural development is widespread. This list of other acts and schemes should give you some idea of the breadth of government influence.

- Green Belt Act (see this chapter);
- IDCs (Industrial Development Certificates) (see chapter 11);
- ODPs (Office Development Permits) (see chapter 11);
- Urban Development Corporations (see chapter 12);
- Enterprise Zones (see chapter 12).

## THE CENTRAL BUSINESS DISTRICT (CBD)

This is the heart of the urban area. It is generally the focus of the transport system and has the greatest accessibility from all areas of the city. Land is limited and expensive here, so developments are multi-storey. The resident population is low.

The CBD also refers to the mix of retailing, commercial and professional offices, city administration and entertainment districts. The larger the town or city, the more distinct is each area within the CBD – therefore these 'zones' might be seen as the standard ones, although in major retailing centres the chain stores, specialist shops, high quality retailing, etc. will separate into distinct **functional areas**, e.g. Oxford Street, Bond Street, Carnaby Street and Saville Row in London's West End. This is because of the **external economies** to be gained by locating close to similar uses.

CBDs can be delineated by:

1 **Mapping land uses** and noting zones of changing use. Uses change towards the edge of the CBD.

2 Using the **central business eight index**, i.e. dividing the total business floor space in a street block by the total ground floor space: over 1 = CBD.

3 Finding the **central business intensity index**: the percentage of floor space in CBD activities related to the area of all floor space. If it is over 50 per cent, then that area is within the CBD.

Both (2) and (3) are difficult to obtain in the UK.

4 Using the **rateable value per metre of street frontage**. This takes the rateable value from the rates books and divides that value by the frontage of the buildings, to give values in £/metre.

5 **Physical features** acting as a barrier to growth, e.g. a hill or a river.

6 **Man-made obstacles to spread**, such as a park, a railway track or a motorway.

7 **An historic quarter** belonging to a city which has been conserved, and which can only be adapted for certain specified functions.

The **core-frame concept** develops the concept of a CBD to take account of a city's dynamism. The **core** is the traditional CBD noted above, whereas uses in the **frame** are characterised by the land uses shown in Fig. 11.11. Very often the CBD is expanding into one part of the frame as a **zone of assimilation** while elsewhere CBD uses are being abandoned, for example losing an area to warehousing and parking lots – a **zone of discard**.

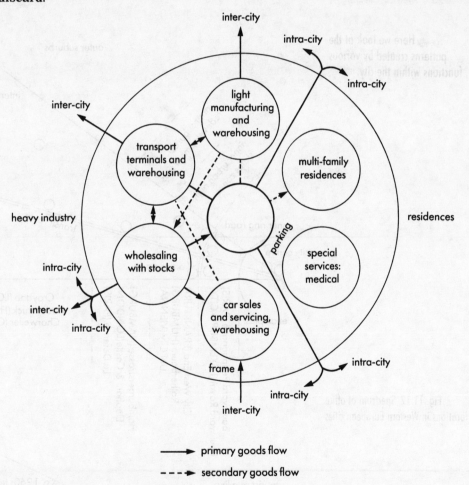

Fig. 11.11 The core–frame concept

→ primary goods flow
--→ secondary goods flow

## PROCESSES IN THE URBAN CORE

- **Comprehensive redevelopment** – the clearance of a site for complete rebuilding, including a new street pattern. Can lead to conflict between interest groups of residents, developers and the council. Les Halles, Paris, and Broadgate Square, London, are two such schemes.

- **Decentralisation of functions** – examples include: (1) offices to the suburbs (Croydon) or to green field sites (M4 corridor) or to other towns (Portsmouth, Bristol etc.); (2) wholesaling – Covent Garden to Nine Elms; Les Halles, Paris, to Rungis; (3) retailing to new peripheral centres – shops in San José, California, all relocated to Eastridge Mall.

- **Tourist district growth**   There may, for instance, be more hotels to cater for luxury tourism, conference tourism and business, e.g. between 1986 and 1989 in Cologne, the Hyatt, the Holiday Inn Plaza, the Intercontinental and the Maritime hotels all opened. Or again there could be purpose-planned tourism districts, e.g. Zeedijk, Amsterdam, and Cologne.

- **Pedestrianisation**   Examples include Above Bar in Southampton, and traffic management in Central Guildford and Coventry.

- **Conservation of historic cores**   For example, Grey Street, Newcastle. Reinforced by acts of parliament, this can lead to gentrification.

- **Gentrification** – the movement of higher socio-economic groups into an area.

### THE PATTERNS OF FUNCTIONS IN A CITY

1   Offices (Fig. 11.12);
2   Retailing (Fig. 11.13);
3   Hotels (Fig. 11.14);
4   Industry (Fig. 11.15).

❝ Here we look at the patterns created by various functions within the city. ❞

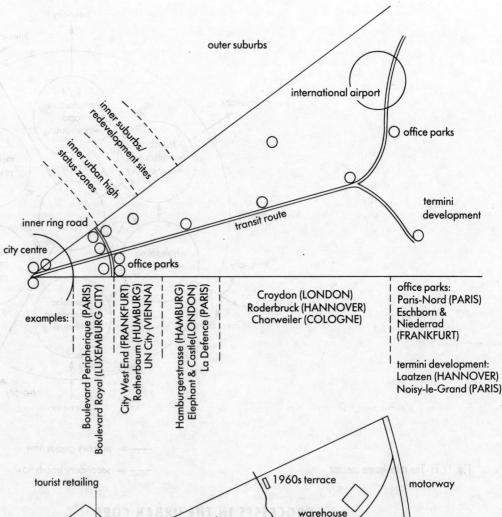

Fig. 11.12 Spectrum of office locations in Western European cities

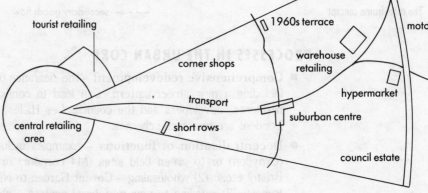

Fig. 11.13 Retail locations in a city

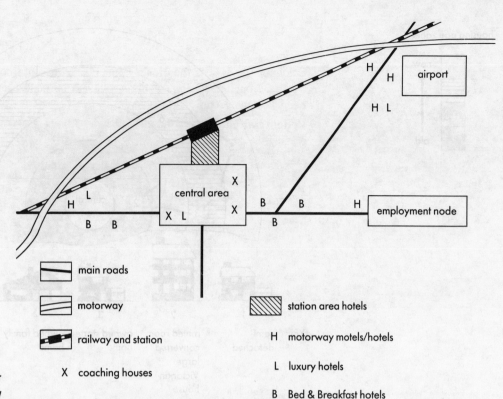

Fig. 11.14 A locational typology for hotels in a European city

**Legend:**
- main roads
- motorway
- railway and station
- X  coaching houses
- station area hotels
- H  motorway motels/hotels
- L  luxury hotels
- B  Bed & Breakfast hotels

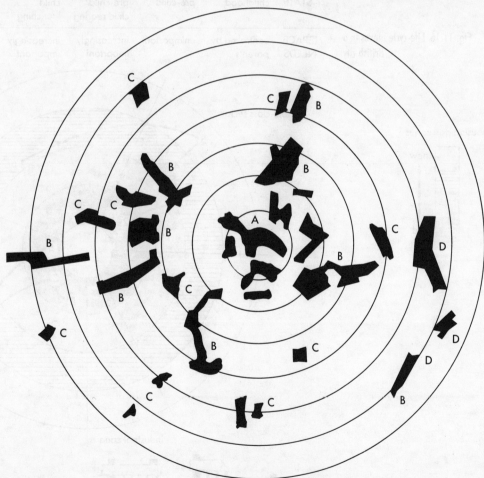

Fig. 11.15 A model of the spatial industrial structure of a metropolis
*Source*: F. E. Hamilton, (1967). This is based on a map of industrial areas in London. The four categories shown are: A = central locations; B = port locations; C = radial or ring transport artery locations; D = suburban locations.

## HOUSING

Housing is one of the main uses of space in cities. The pattern of housing choice (or even non-choice) is based upon (1) life-cycle and (2) social class or life level. Throughout our lives our housing needs change (Fig. 11.16). If social group life chances are added to this, we get the contrasting patterns of housing choice through time (Fig. 11.17).

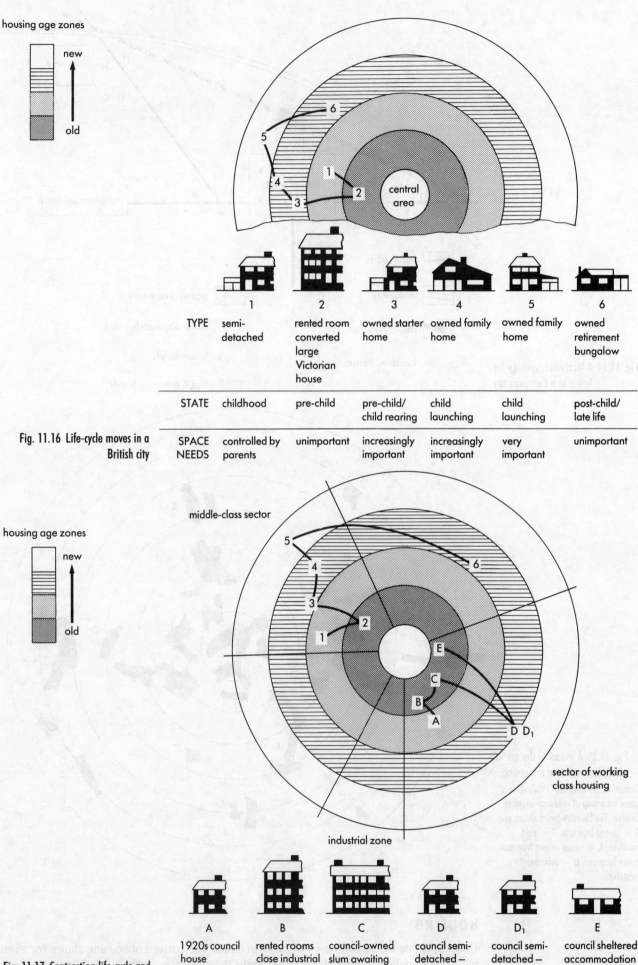

housing age zones

new

old

TYPE

| | 1 | 2 | 3 | 4 | 5 | 6 |
|---|---|---|---|---|---|---|
| TYPE | semi-detached | rented room converted large Victorian house | owned starter home | owned family home | owned family home | owned retirement bungalow |
| STATE | childhood | pre-child | pre-child/child rearing | child launching | child launching | post-child/late life |
| SPACE NEEDS | controlled by parents | unimportant | increasingly important | increasingly important | very important | unimportant |

Fig. 11.16 Life-cycle moves in a British city

middle-class sector

housing age zones

new

old

sector of working class housing

industrial zone

Fig. 11.17 Contrasting life-cycle and social housing moves in a British city

| A | B | C | D | D₁ | E |
|---|---|---|---|---|---|
| 1920s council house | rented rooms close industrial area (compulsory purchase) | council-owned slum awaiting demolition – married | council semi-detached – family home | council semi-detached – purchased after 'n' years renting | council sheltered accommodation for elderly |

Housing is also subject to **filtering** (Fig. 11.18) in which housing tends to move 'downwards' through social groups, although in some areas the process is reversed by **gentrification**.

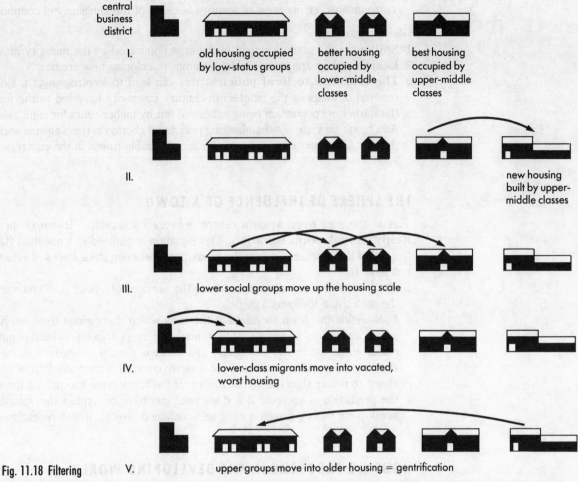

I.

central business district

old housing occupied by low-status groups

better housing occupied by lower-middle classes

best housing occupied by upper-middle classes

II.

new housing built by upper-middle classes

III.   lower social groups move up the housing scale

IV.   lower-class migrants move into vacated, worst housing

V.   upper groups move into older housing = gentrification

Fig. 11.18 Filtering

Choice patterns are constrained for lower income groups, for students and for those who do not fit into the usual pattern of life in our society. Students can often outbid the poor by gaining access to moderate-sized houses and using all the rooms as bedsits, so reducing per capita costs by more than a family of four with a sole wage-earner.

Housing is a series of markets for each type of house. The choice is constrained by 'gatekeepers', including the institutions or their representatives who approve finance, the estate agents, building societies, landlords, the housing departments of local authorities, and so on. Building societies used to **red line** certain areas as unsuitable for loans – many inner cities were treated like that. Other inner areas were **green lined**, i.e. approved because of factors such as gentrification, future potential or possessing the 'right address'. Certain areas acquire **positive externalities** – a park, sea view, hillside or the medieval heart of a town – while other areas have **negative externalities** – high crime rates which label particular areas, poor schools and a lack of facilities.

## SOCIAL SEGREGATION: THE CASE OF THE GHETTO

**Ghetto** was the name of an area in Venice where the Jews were confined, and its name became associated with an area set aside for the Jewish community. Today it can refer to an area where there is an over-concentration of any minority group: blacks in the USA, Turks in West Germany, Indians, Pakistanis and Bangladeshis in Great Britain. Ethnic residential clusters form in response to four functions of clustering:

a) **Avoidance**, by focusing inward the religious and cultural needs, e.g. the mosque or temple and eating Hallal meats. Such customs are best 'developed' within a close community.

b) **Preservation of culture** especially in the home and through distinct social organisation.

c) The **attack** function is frequently noted because of the more violent manifestations of ethnic clustering seen in riots. But there are peaceful attack functions, such as the pressure to form an immigrant worker assembly in the city of Cologne.

d) The **defensive** function helps new immigrants because they move into similar surroundings, giving greater security because of the familiarity of continuing customs.

Ghettos grow by:

1 **Spillover**: the gradual spread of the area dominated by the minority group;
2 **Leapfrogging**: the more sudden attempt to colonise new areas;
3 **The response to local policies**: this can lead to leapfrogging. In Cologne, urban renewal of areas of the nineteenth-century inner city have led to the foreign worker (Gastarbeiter) population being squeezed out by higher rents for refurbished property. As a result they are now establishing peripheral ghettos in the high-rise social housing on the city fringe and even out into the less desirable towns in the city region.

## THE SPHERE OF INFLUENCE OF A TOWN

This is the area over which a centre delivers its services. It can be predicted using **Reilly's break-point equation**. This equation is outlined in a question (Question 5) at the end of the chapter. To map the break point between the spheres of influence of towns you have to:

1 Study where people come from to use the services of a place, i.e. you must go out into the area using Reilly as a guide,
2 Look inside the town to establish the area which the various town services *expect* to serve; deliveries, advertisements, school catchment areas, provision of public services, labour force catchment areas, etc. can all give a guide. However, none of the lines showing a particular catchment will exactly coincide – they will form a **zone**, which is closer to reality than the line of theory. If the break-point line passes through the zone the prediction is correct. If it does not, you have to explain the variations. Testing break-point theory makes a good self-contained project, if well researched.

## HOUSING PROBLEMS IN THE DEVELOPING WORLD

Rapid urbanisation has led to **spontaneous settlements – favela, barriadas, bustee** and **shanty town** are alternative terms used in some countries. These **squatter settlements** appear quickly as rudimentary homes with no sanitation, water supply, power, medical or educational facilities. At the worst they are a blanket in the gutter, a concrete pipe or a shack on a roof. As the infrastructure is developed and as some attain better wages, so the dwellings become more permanent 'slums of hope'. Other areas decline still further. Governments have the unenviable task of improving such housing situations by:

a) Developing an economy that can support the clearance of the worst areas: in Hong Kong, the 'boat people' from Vietnam are housed in a new version of the spontaneous settlement, the **refugee camp**.
b) Building **new towns**, e.g. Yishun, Singapore, and Cuidad Guyana, Venezuela.
c) Providing **site and service schemes**, as in Zambia where the state provides the foundations and basic facilities and permits the occupier to build as and when improvements can be afforded.

**URBAN POLICY ISSUES**

## NEW TOWNS

These are normally comprehensively planned settlements, developed over a short period of time. There are often governmental policy reasons for their development. In the UK they were developed after the **1946 New Towns Act** although the origin of the pressure for new towns was in the **Garden City Movement**. In Great Britain they were developed to:

Reasons for New Towns.

a) Provide settlements to accommodate overspill – e.g. eight around London;
b) Revive depressed regions (see chapter 13), e.g. the new towns of the North-East;
c) Aid the development of a resource, e.g. Corby (steelworks);

d) Expand large urban centres, e.g. Peterborough;

e) Create a counter-magnet to London, e.g. Milton Keynes.

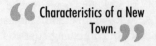

**Characteristics of a New Town.**

Over time, the size of the towns has grown to provide the higher threshold population for a wider range of services.

Features of the new towns' townscape:

1 New central area with pedestrianised shopping area;
2 Industry and offices in distinct zones;
3 Pedestrian and vehicular segregation – some have segregated public transport, e.g. Runcorn;
4 Use of the **neighbourhood unit**: the provision of low-order services to an ideal, small-scale, community whose limits were defined by the major road network and green wedges – they contain all the facilities thought to be necessary at that time for a full community life.

New towns have been developed in other countries:

- In France, to structure the suburbs of Paris, e.g. Marne-la-Vallée and St Quentin en Yvelines, or to accommodate overspill e.g. l'Ile d'Abbeau, Lyons.

- In the Netherlands, to provide service centres on the polders, e.g. Emelloord (on the polder where Christaller influenced the pattern of settlement) and Lelystad, or to cater for overspill, e.g. Almere.

- In the NICs (chapter 13) they are to cater for rapid urbanisation, e.g. Tsuen Wan in Hong Kong, and Yishun and Jurong in Singapore.

- In socialist countries they are built to keep the size of cities manageable, e.g. Halle Neustadt, or to exploit a resource, e.g. Hoyerswerda; or for a major industrial complex, e.g. Eisenhüttenstadt – all in the German Democratic Republic.

## GREEN BELTS

These are established:

a) To check the growth of an urban area;   b) To prevent towns merging;   c) To preserve the special character of a town;   d) To provide for recreational needs.

In Britain the **Green Belt Act 1938** began the process. There are six types of Green Belt (Fig. 11.19):

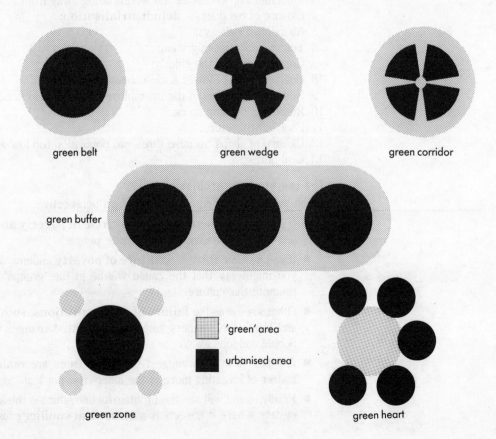

green belt          green wedge          green corridor

green buffer

'green' area

urbanised area

green zone          green heart

Fig. 11.19 A classification of 'green' planning

**"Types of Green Belt."**

1 Green Belt, e.g. London;
2 Green Wedge, e.g. Copenhagen;
3 Green Corridors, e.g. Geneva;
4 Green Buffer, e.g. Ruhr area;
5 Green Zone, e.g. Paris;
6 Green Heart, e.g. Netherlands.

**"Problems of Green Belt."**

### The problems of Green Belts

■ They encourage development to **leapfrog** the Green Belt to towns beyond.

■ **Infilling** and development restrictions lead to high-cost housing.

■ The areas are encroached upon by mineral workings, hospitals and sports grounds – all permitted.

■ Transport routes take advantage of the lack of development, e.g. M25 and Ruhr motorways.

■ Developers want land here for housing overspill, e.g. Foxley Wood.

■ They exclude the lower socio-economic groups and the young as market forces push up property prices.

## THE INNER CITY PROBLEM

This is probably the largest and most complex urban issue. It has links to the economic geography of the country (chapter 10) and to aspects of population change and migration (chapter 3), but its location is specifically city-based. It is *not* solely a UK problem and therefore you might use examples from other countries: USA, France, West Germany, which all have inner city problems.

It is a long-term issue – when did Burgess note the problems in inner Chicago? Prime Minister Wilson announced an Urban Programme before almost all of the readers of this book were born! Has much happened in your lifetime and if so how, why, where, and with what consequences?

**"Evidence of an inner city problem."**

### The evidence for the problem

1 Factory closures in the inner city – **deindustrialisation**;
2 Factory relocation from inner cities – **deindustrialisation**;
3 Movement of wholesale and warehousing away from inner city;
4 Closure of old docks – **deindustrialisation**;
5 Abandoned railways;
6 Young family out-migration;
7 Concentration of the elderly;
8 Concentration of lower socio-economic groups;
9 Higher proportions of the unemployed and those in need of state assistance;
10 Run-down building stock;
11 Poor infrastructure;
12 Closure of shops because threshold population too low and too poor;
13 Civil unrest.

### The causes of the problem

Much will depend of course on your own perspective:

■ If you feel that there is a **vicious circle of poverty and deprivation**, then you will see the cause as one of 'inadequate people'.

■ If you feel that there is a **culture of poverty** endemic in society in these areas, then you might say that the cause will lie in the 'groups' (gangs and societies) which promote this culture.

■ Others see it as the **failure of our institutions**, such as those involved in planning and blame the planners who have permitted changes in the outer city which suck people and jobs away.

■ Another view is to suggest that **resources** are **maldistributed** and that it is a matter of investing more in the inner city, not less, as has often happened.

■ Finally, some will see the problem as one which is the spatial expression of the class system – here it is seen as a **structural conflict** caused by class divisions.

### The reactions to the problem

The Urban Programme announced in 1968 has continued for over 20 years and is now larger and more widespread in its coverage of cities.

| PROGRAMME | EXISTING NUMBER | EXAMPLES |
| --- | --- | --- |
| Partnership Authorities | 7 | Hackney<br>Liverpool |
| Programme Authorities | 31 | Bolton<br>Wolverhampton |
| Designated Districts | 15 | Newham<br>Barnsley |
| Garden Festivals | 5 | Liverpool<br>S. Wales 1992 |
| Enterprise Zones | 24 | Swansea<br>Corby<br>Isle of Dogs |
| Urban Renewal Grant (City Grant) | 5 | Salford<br>Nottingham |
| Urban Development Corporations | 12 | London Docklands<br>Lower Don |
| City Task Forces | 7 | Liverpool<br>Teeside |

Table 11.1 Urban programmes

For each of these programmes you might need to know the following:

1 Where has it occurred?
2 For how long? For example, new Enterprise Zones are no longer being designated although the policy will last until the mid-1990s.
3 What were the effects on the designated area and on the areas surrounding it? Newspaper cuttings will help here. How has employment and the built environment changed (see chapter 12 as well)?
4 Who has gained and who has lost from the changes?

## RURAL SETTLEMENT

## FACTORS INFLUENCING THE SITES OF VILLAGES

a) Water supply – spring lines – especially in the past;
b) Route focus;
c) Defensive needs – past, and present in some countries;
d) Access to suitable lands – especially in medieval period;
e) Distance from other villages, access to common lands, (overlaps with (d);
f) Planning policy when new villages developed.

## THE FORM OF VILLAGES

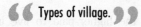 Types of village.

1 **Linear**  a) along a valley
            b) along a dyke
            c) along a road.

2 **Nucleated**  a) at a natural focus
               b) at a junction
               c) at a change of transport mode
               d) at a defensive point.

Nucleated can take various shapes, of which the most recognisable today is the **green village**, a settlement surrounding a round, linear or oval green, common in North-East England. Most villages today are nucleated because planning controls have prevented the spread of villages into the countryside. **Infilling** of empty sites is preferred to **accretion**. Nevertheless, many villages are now very large.

## DISPERSED SETTLEMENT

This is a pattern of isolated farms and hamlets. It has come about because of:

a) The type of **inheritance** – *Gavelkind,* by which land was inherited by all the sons (by daughters only if there were no sons!), led to dispersed patterns in, for example, Central Wales;

b) The effects of **parliamentary enclosure** which resulted in some farms moving out of the village to be on their new holding;

c) The effects of remembrement (see chapter 10) in France and West Germany which was similar to enclosure, although it occurred in the period 1950–80;

d) The consequences of a planned rural settlement policy, e.g. the Ijssel Meer polder where hamlets of adjacent farms were spread across the landscape;

e) Government policies to reclaim marshes, alter landholdings and reorganise areas, e.g. Pontine marshes in Italy.

### Measuring dispersion

This can be attempted using **nearest neighbour analysis** (chapter 2 includes a question which has used this technique). The nearest neighbour statistic or ranges from 0.0 = **clustered**, through 0.23 = **linear clustering**, 1.0 = **random**, to 2.15 = a **regular pattern** (Fig. 11.20).

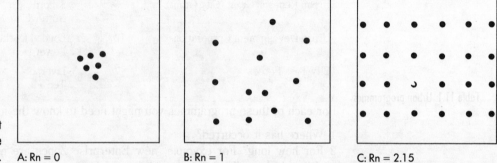

clustered  random  regular

A: Rn = 0  B: Rn = 1  C: Rn = 2.15

Fig. 11.20 Hypothetical settlement patterns. Rn values indicate degree of randomness.

■ **Stage 1** – Define the area to be studied and locate the pattern of settlements in it. The dots on the edge should either be excluded, in case they are close to others outside the study area, or used only when they are closer to dots in the study area.

■ **Stage 2** – Number all the dots and calculate the distance to each nearest neighbour. Add these up to get the **observed mean** = $\bar{D}_o$. The density of points is the number of points ÷ area of the study.

■ **Stage 3** – Calculate the expected means for a random distribution:

$$\bar{r}\,E = \frac{1}{2\sqrt{\text{density}}}$$

■ **Stage 4** – Now you can calculate: $Rn = \dfrac{\bar{D}_o}{\bar{r}E}$

Once you have your $R_n$ value you must then ask why the pattern is as it is.

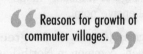

## COMMUTER VILLAGES

This was the term devised by Pahl to note the growth in population in the villages of Hertfordshire. Numbers of people in villages grew because of:

a) The decentralisation of industry to new towns, enabling the higher-paid to live in villages;

b) Greater affluence – more money to spend on transport/commuting;

c) Improved public transport – electrified rail lines – back into London;

d) Policies which expanded some villages and enabled developers to build for the new 'villagers';

e) Perceptions of village life which attracted people;

f) Green Belt policies which restricted growth of some villages. This made those villages more desirable, so that house prices rose, forcing development beyond the Green Belt.

The consequences of the rise of commuter villages were:

a) A range of new social classes in the villages;

b) These newcomers were often segregated into a separate area;

c) A progressive reduction in the availability of homes for those whose employment and family were actually in the rural area;

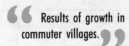
**Results of growth in commuter villages.**

d) The focusing of housing for the lower-paid in certain villages where small council house estates were developed;

e) The reduction in 'low order' goods provision, as the new villagers shopped in the town;

f) An increase in services for the more affluent: restaurants, antique shops, etc.;

g) A decline in public transport (because more cars are used), which particularly affects the old, the young and the poor.

**NB** This process continues as **counter-urbanisation** and increases the dominance of an urban way of life, even in remote rural areas.

## SECOND HOMES

These are dwellings, normally, but not exclusively, in the countryside whose residents live elsewhere. Many were holiday retreats and often not very substantial. However, *rural depopulation* and farm abandonment has resulted in many farms, barns and other rural buildings being taken over and redeveloped as second homes. The Massif Central, France, and Central and North Wales are just two areas where second homes are very common. What are the consequences?

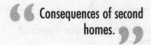
**Consequences of second homes.**

a) The improved upkeep of villages, hamlets – less dereliction;

b) Local employment in improving and maintaining property;

c) An asset for the owner which can earn further income;

d) A rise in rural house prices, often beyond the means of, for example, farm workers;

e) The provoking of opposition – arson attacks;

f) An alteration in the age structure of the settlement.

Some second homes are purpose built, e.g. the coastal settlements of Languedoc or the time-share apartments in the Lake District and at Aviemore.

## RURAL DEPOPULATION

This is a form of population migration (chapter 3) although it can be caused by **high mortality**. The *causes* are:

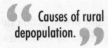
**Causes of rural depopulation.**

a) Loss of jobs in agriculture;

b) Mechanisation of agriculture;

c) The perceived benefits of urban life;

d) Housing conditions in the rural area;

e) The perceived lack of provision of rural services;

f) Marriage, leading to a move to the partner's home;

g) In the Developing World, labour migration to the towns.

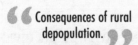
**Consequences of rural depopulation.**

The *consequences* are

a) Abandoned property and sometimes land left fallow;

b) A greater potential for second homes;

c) The loss of services and public transport in rural areas as threshold numbers decline;

d) The loss of social cohesion in the rural community;

e) An ageing population, because it is usually the young who leave, often taking their families as well;

f) Greater distance for villagers to travel in order to obtain goods and services;

g) In the Developing World, the loss of some labour to towns, although the impact may not be that noticeable.

It is probably best to distinguish rural depopulation in terms of its impact on the Developed and Developing Worlds. Can you do this?

# EXAMINATION QUESTIONS

### Question 1 The challenge of urbanisation

Study Fig. 11.21 which shows the ethnic composition and the worst areas for certain social and economic conditions in Fresno, USA.

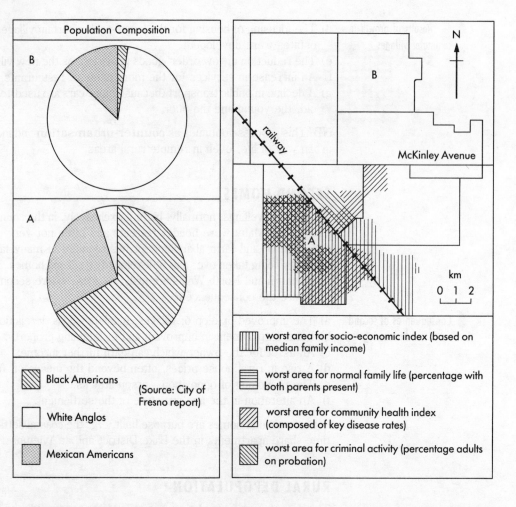

Fig. 11.21

a) Critically assess the criteria used to identify, and techniques used to map, the locations of the worst areas. (8)

b) In a survey of Fresno, residents were asked to identify the major urban problems and these are shown in rank order in Table 11.2. The rank orders given by residents in two neighbourhoods (A and B on the map) are also shown. Suggest reasons for the variations in ranking. (9)

c) With reference to examples of cities in developed countries you have studied, explain why the variations in social and economic conditions present a challenge for urban decision-makers. (8)

| OVERALL RANK | THE TEN HIGHEST-RANKED PROBLEMS IN THE CITY-WIDE SURVEY | RANK GIVEN IN AREA A | RANK GIVEN IN AREA B |
|---|---|---|---|
| 1 | Health services cost too much | 1 | 1 |
| 2 | Too much drug abuse | 2 | 2 |
| 3 | Too much flooding when it rains | 21 | 3 |
| 4 | Too many people get welfare who should not get it | 29 | 4 |
| 5 | Not enough jobs | 3 | 15 |
| 6 | Not enough safety from crime | 24 | 7 |
| 7 | Not enough litter clean-ups | 20 | 6 |
| 8 | Not enough money spent on basic education | 5 | 13 |
| 9 | Housing too expensive to rent/buy | 6 | 12 |
| 10 | Too much air pollution | 41 | 5 |

Table 11.2

(London 16–19, June 1989)

## Question 2
Study the map (Fig. 11.22) showing world urbanisation by country.

a)  i)  Give a short definition of urbanisation. (2)
    ii)  Identify THREE main characteristics of urbanisation. (3)

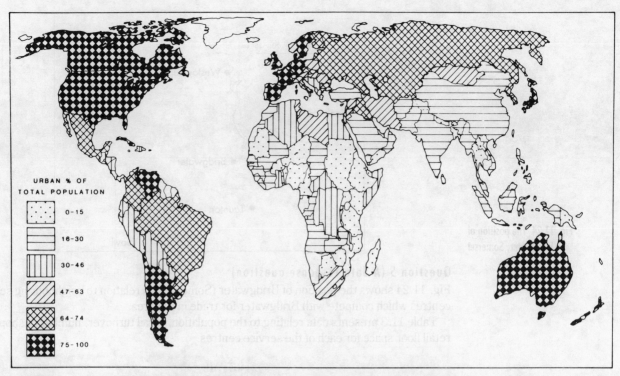

URBAN % OF
TOTAL POPULATION

0 – 15

16 – 30

30 – 46

47 – 63

64 – 74

75 – 100

**Fig. 11.22**

b) i) Briefly account for the high degree of urbanisation in ONE of the named world regions where there is generally more than 75% of the population in urban areas. *(4)*

ii) Briefly account for the low degree of urbanisation in ONE of the named world regions where there is generally less than 25% of the population in urban areas. *(3)*

c) Give reasons in support of the view that in some parts of the world the degree of urbanisation might be:
   i) greater than indicated by the map; *(4)*
   ii) less than indicated by the map. *(4)*

## Question 3

a) What is dispersed settlement? *(3)*

b) Name TWO techniques which may be used to measure the spatial dispersion of settlements. *(2)*

c) Identify the TWO types of dispersion shown by the diagrams below. *(2)*

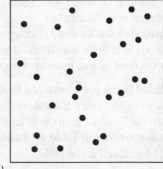

**Fig. 11.23a)**

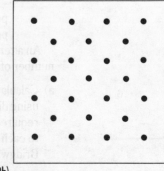

**Fig. 11.23b)**

d) For ONE named area of dispersed settlement, identify the physical and human factors which may have encouraged that dispersion. *(8)*

(AEB, June 1989)

## Question 4

a) Outline the basic assumptions and principles upon which Christaller based his theory of central places. *(15)*

b) Suggest reasons why the model has limited application today. *(10)*

(AEB, June 1988)

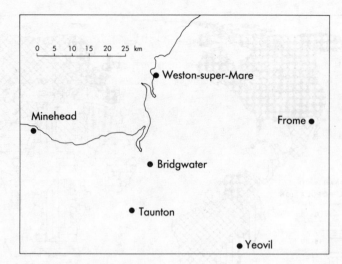

Fig. 11.24 The position of
Bridgwater, Somerset

## Question 5 (A data response question)

Fig. 11.24 shows the position of Bridgwater (Somerset) in relation to neighbouring service centres which compete with Bridgwater for trade in the area.

Table 11.3 presents data relating to the population, retail turnover, number of shops and retail floor space for each of the service centres.

| CENTRE | POPULATION | RETAIL TURNOVER £000 | NUMBER OF SHOPS | RETAIL FLOOR SPACE (m²) |
|---|---|---|---|---|
| Bridgwater | 26 598 | 10 661 | 305 | 21 800 |
| Weston-super-Mare | 50 794 | 21 052 | 646 | 45 900 |
| Frome | 13 384 | 4 749 | 205 | 11 200 |
| Yeovil | 25 492 | 14 439 | 310 | 29 500 |
| Taunton | 37 373 | 21 658 | 456 | 53 900 |
| Minehead | 8 063 | 4 620 | 145 | 12 400 |

Table 11.3 Population and retail data for selected towns

The sphere of influence of a central place can be predicted by using Reilly's break point equation.

$$d_{jk} = \frac{d_{ij}}{1 + \sqrt{\dfrac{\text{pop } i}{\text{pop } j}}}$$

where  $d_{j}k$ = distance of break point k from settlement j (Bridgwater in this exercise)
$d_{ij}$ = total distance between settlements i and j
pop i = population of settlement i
pop j = population of settlement J (Bridgwater).

An alternative method could be to substitute other indicators such as retail turnover, number of shops or retail floor space, instead of population.

a) Calculate the predicted break point between Bridgwater and neighbouring settlements using different indicators and enter the values (to 1 decimal place) in Table 11.4. You are required to calculate the 7 values omitted from the table. The distance from Bridgwater to each settlement is shown in on Table 11.4 and the distance of the break points from Bridgwater should be entered in the table. *(7)*

| DISTANCE FROM BRIDGWATER TO SETTLEMENT | | POPULATION BREAK POINT | RETAIL TURNOVER BREAK POINT | NUMBER OF SHOPS BREAK POINT | RETAIL FLOOR SPACE BREAK POINT |
|---|---|---|---|---|---|
| Weston | 24 | 10.1 | 10.0 | | |
| Frome | 46 | | 27.5 | 25.3 | 26.7 |
| Yeovil | 32 | | 14.8 | 16.0 | 14.8 |
| Taunton | 13 | 5.9 | | 5.9 | |
| Minehead | 34 | 21.9 | 20.5 | 20.1 | |

Table 11.4 Predicted break-point distances from Bridgwater to neighbouring settlements

All break-point distances should be given from Bridgwater.

b) Using Fig. 11.24 as a base, draw a map to show the predicted sphere of influence based upon population totals for the settlements. *(3)*

c) On one side of arithmetic graph paper draw two scattergraphs:
   i) to show the relationship between retail turnover and number of shops. You should plot retail turnover on the vertical (y) axis and identify each point by the first letter of the settlement name;
   ii) to show the relationship between retail floor space and population. You should plot retail floor space on the vertical (y) axis and identify each point by the first letter of the settlement name. *(4)*

d) By referring to the data available, comment on the variations in the break-point distances between Bridgwater and the other settlements as indicators of the characteristics of each of the service centres. *(6)*

(JMB, 1988)

### Question 6
Discuss the processes which lead to the development of distinct social areas within cities. *(25)*

(London 16–19, 1989)

### Question 7
a) Discuss the factors which give rise to linear forms of rural settlement. *(15)*

b) In what ways might these factors differ from those encouraging linear extensions to urban areas? *(10)*

(AEB, Nov 1988)

### Question 8
Examine the effects of counter-urbanisation on the economic and social geography of rural settlements. *(25)*

(AEB, June 1989)

### Question 9
This is a multi-stimulus question involving photographs and an Ordnance Survey map. You should be able to relate the photographs to the map in part (b). Part (a) draws in part on the material of chapter 10.

Study the Ordnance Survey map extract provided (Tyneside 1:50 000 Landranger Series Sheet 88).

a) Compare and contrast the nature of primary and secondary industry in the areas west and east of the A1(M) road. *(12)*

b) Using the map extract overleaf and Figs 11.25 and 11.26, photographs of Chester-le-Street and Washington, describe and explain the differences in form and function between the two settlements. *(13)*

(AEB, 1984)

**Fig. 11.25**          **Fig. 11.26**

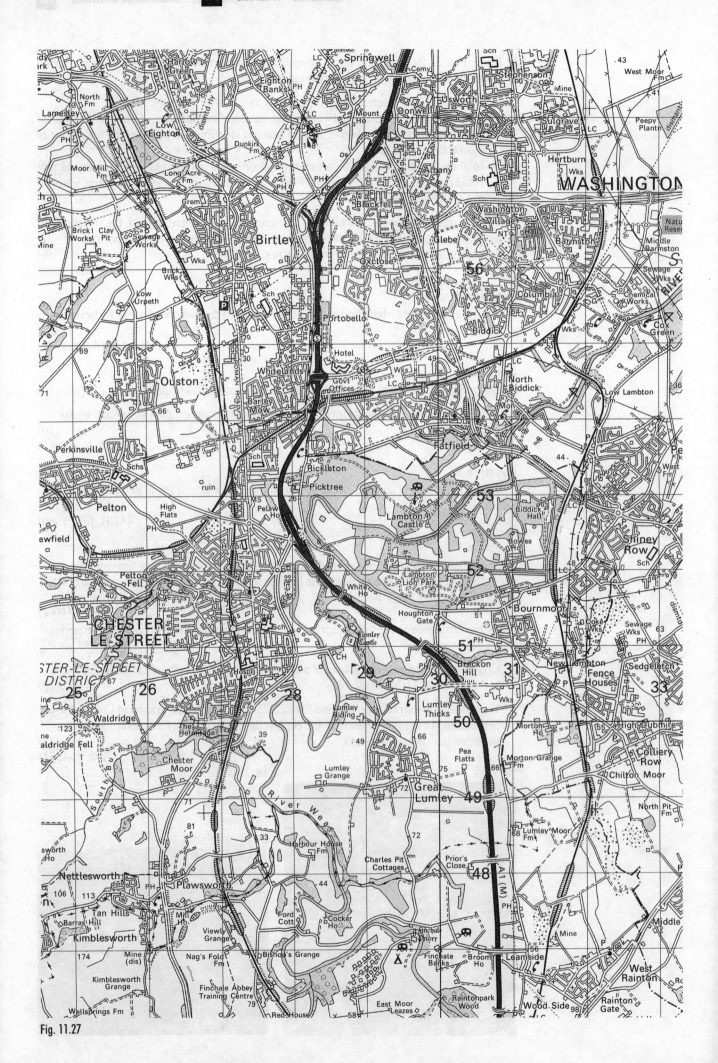

Fig. 11.27

# OUTLINE ANSWER

### Answer to Question 1   The challenge of urbanisation

a) The map and pie graphs of Fresno do begin to show some aspects of well-being, but you could point out that they miss certain aspects, such as the age, type and quality of the housing. The map is based on percentage data but you are not told what 'the worst' is, is it 5 per cent or 70 per cent? The data makes assumptions about normal life, e.g. two parents resident is normal, but to whom? The diseases are not specific so could include diseases of wealth. The map is a choropleth achieved by 'sieve technique' – each layer adds to the darker shade but it leaves the boundaries crude. The pie charts are part of Fig. 11.21 and the sizes are equal but that does not mean that the population in each area is equal! It could be that there are more Mexicans in B than in A.

b) The columns are the rank positions with more problems omitted. It shows great variability in perception of problems. Variations might be based (there is no proof other than the map) on wealth, income, health, crime and family status, some of which are variables in social-area analysis.

   Area A is worried about housing, employment and service provision whereas B's residents are more concerned by criminality and the environment.

   Flooding is an odd-ball – if you know where Fresno is you might realise that flash-floods can occur. There is an implied ethnic dimension, but the sampling frame does not help you here. Nevertheless, the main problems are identical in the perception of all residents.

c) It is about developed and *not* developing countries. It would be best to select a city, outline some variations and then say why they are a challenge. Perhaps you could take part of a city, such as London's East End or East Paris. Try then to focus on how urban decay and deprivation are challenges – perhaps because the areas involved present different opportunities for different groups. We might feel that the decision-makers should decide in the best interests of all. However, there is obviously a political dimension here: do the decision-makers represent one view or ideology – perhaps that carrying most votes! If so they may do little to alleviate the challenging problems!

## STUDENT ANSWER WITH EXAMINER'S COMMENTS

### Question 3

> **You have not stated what the pattern is. It needs to say rural.**

> **This is correct.**

> **No – you need quadrat analysis.**

> **Both correct.**

> **The right theme but no reasoning.**

> **You never say how these encourage dispersion.**

a)   Settlement with no semblance of order. It describes the layout between different settlements which interact.

b)   i) Nearest neighbour analysis.
     ii) Centre of gravity.

c)   A) Random   B) Uniform or Regular

d)   In Highland Scotland, physical barriers include mountains, rivers and climate. Low lying areas are favourite for dispersed settlements.
     Transport links are important between settlements. Because settlements are mostly agricultural, soil fertility, relief and markets had an important part to play.

# REGIONAL DEVELOPMENT AND PLANNING

## GETTING STARTED

Much of your revision should get easier from now on! In this chapter, and the two which follow, there are references back to concepts, theories and principles which have been noted elsewhere. Nevertheless, there are new theories to be revised.

As we saw in chapters 1 and 2, the study of the region was once regarded as the summit of geographical enterprise, involving as it did the correlating and relating of a disparate set of physical and human facts. Today, the integration of explanatory variables is used more to explain *why* an area or place functions in a particular way and also to suggest *how* it might function better in the future. Emphasis has been placed on spatial inequalities in life chances, on economic development and on the factors causing uneven rates of development. **Planning**, whether regional or physical, is our attempt to rectify these imbalances and inequalities.

If you study JMB 'C' then this chapter is important for those regions in the Developed World which you have studied. In other syllabuses, regional development can be covered as a part of a specific integrating section (e.g. AEB, London) or as part and parcel of sections on economic and settlement geography. In other words, regional development appears in some form in *all* syllabuses.

Planning is also present in almost all syllabuses and particularly in those which examine current issues (e.g. London 16–19, AEB, NISEC and WJEC). Planning is probably also going to feature in many projects (see chapter 15), either through your study of the consequences of planning or when you examine future changes in a landscape or townscape.

**BASIC THEORIES IN REGIONAL DEVELOPMENT**

# ESSENTIAL PRINCIPLES

### Economic Base Theory or Export Base Theory

This is a useful starting point. Any area has two types of economic activity:

- **Basic** (or **export base** activities) which are founded upon demand originating from *outside* the area. These industries' **location quotient** – a coefficient which compares the percentage share of employment in an activity within an area with its percentage share nationally – is greater than 1·0.

- **Non-basic** (or **residentiary activities**) are those activities whose product is used within the area and which are dependent on the area's **internal** markets. The non-basic function can be subdivided into **consumer-serving** activities and **producer-serving** activities.

An increase in the basic activity of a region will bring in extra income. This can then be used in a variety of ways:

a) To increase the demand within the region for goods and services;
b) To invest in further growth of the basic activity.

Basic activity also has a **multiplier effect**, i.e. it increases economic activity throughout the region. It is normally expressed as an **employment multiplier**, indicating the number of extra jobs created as a result of the direct and indirect expansion of an activity. If you look at the growth of the economies in, for example, North-East England or in South Wales during the nineteenth century, the role of the basic activities of coal, steel and shipbuilding are evident.

### Inter-regional trade multipliers

Inter-regional input–output analysis is a complex analysis which examines transactions *between regions* as well as *between the sectors of the economy within the region*. It can be used to measure the impact of the planned development of, say, a car plant or of increased tourism on a region. The actual analysis is complex and is not needed at A-level. However, a knowledge that there are sophisticated models dependent on computers for their operation might be a useful concluding point in an essay.

### Industrial location theory (see chapter 10) and central place theory (see chapter 11)

These are also helpful in explaining the patterns of growth and development within a region. Similarly Myrdal's model of cumulative causation, which was produced to explain global development (and is therefore explained in chapter 14) is also relevant to help explain regional growth and decay.

### Growth pole theory

This was developed by the economist **Perroux** (1964), who noted that economic growth does not appear everywhere at once, but in points or **development poles** with variable intensities, before spreading to the whole economy. **Boudeville** (1966) applied the theory to the region, showing that a set of expanding industries located in an urban area would induce further economic development throughout the region.

- **Leading industries** are the large firms which dominate a growth pole and so generate economic development. In planning a growth pole, these are usually grafted onto an existing urban centre, e.g. Taranto in Italy.

- **Polarisation** is the effect of the rapid growth of other economic activities in the growth pole consequent upon the leading industry's growth.

- **Agglomeration economies** (chapter 10) are implicit in polarisation. The geographical consequence is the concentration of resources and economic activity in a limited number of centres.

- **Spread effects** or **trickling down** are the effects of growth moving out into the region as a result of, for example, demands for food and for other resources to be used in the growth pole (the term was used by Myrdal, see chapter 14).

This theory was used in the 1960s to encourage regional growth in North-East England and Central Scotland. It was also used in the Bari–Brindisi–Taranto triangle in southern Italy and in the French policy of *Mêtropoles d'equibres*. It has also been applied to developments in Brazil, Venezuela, India and the USSR.

'Trickling down' is an attractive policy, because:

**Support for 'trickling down'.**

1 Agglomeration economies are an efficient way of generating growth;
2 The **concentration** of investment involves less public money on, for example, roads and other infrastructure.
3 The spread effect helps the surrounding area.

---

## REGIONAL PLANNING IN GREAT BRITAN

### The identification of regions

Regions can be identified in a number of ways. By:

**Identifying a region.**

1 Changing employment structure, e.g. deindustrialisation;
2 Unemployment levels;
3 Migration, especially out-migration;
4 The strength of the regional economy, leading to the identification of **depressed** and **pressured regions**.

### The case for regional planning

Should we have regional planning? The original arguments involved 'Keynesian economics' and were in favour of government intervention. This would help ensure the maximum utilisation of underused scarce resources in depressed areas and diffuse the over-utilisation of resources in pressurised regions. Politically, investing in regional development can also pay in terms of votes, seats in parliament and, perhaps, in continuing in power as a government. Nationalist politics in Wales and Scotland depend on this appeal, among others. If there *is* a regional policy, then the choices are:

a) To move work to the workers;
b) To move workers to the work ('get on your bike'); or
c) A mixture of (a) and (b).

### Evolution of regional policy in Great Britain

The origins of regional policy are in the Depression years of the 1920s and early 1930s.

**Legislative base for regional policy.**

- **1934 Special Areas Acts**: the North-East, South Wales, West Cumberland and Clydeside/North Lanark;
- **1940 Barlow Commission Report**: the foundation of regional planning legislation to 1979. Proposed measures to assist depressed areas and to decongest South-East and London and New Towns;
- **1945 Distribution of Industry Act**: first major government intervention in peacetime (see chapter 10);
- **1946 New Towns Act**: (see chapter 11).

These were the foundations of policy. At maximum extent of policy (see Fig. 10.16, chapter 10), there were various types of assisted area:

- **Special Development Areas 1967–84**. These had, or were likely to have, very high, persistent unemployment. They received the highest level of aid. Their average level of unemployment in 1983 was 18.6 per cent, compared with a national figure of 13.4 per cent of the working population living in these regions.

- **Development Areas**. These were regions scheduled to receive aid from the government because of their poor economic performance. The policy for these areas was designed to bring industry to the workers. The average unemployment in 1983 for Development Areas was 16.4 per cent of the working population living in these regions. These areas are a continuation of the Depressed Area policy of the 1930s.

- **Intermediate Areas** (1970). The 'Intermediate' or 'Grey' Areas were proposed by the Hunt Committee. They are areas where economic difficulties are clearly apparent, but are not as bad as those in the previous two categories. Selected forms of aid were available to such areas.

The Welsh (WDA) and Scottish (SDA) **Development Agencies** have continued to operate, attracting industry to these regions.

## Policy changes since 1979

The Thatcher government has rolled back the map of regional aid (see Fig. 10.16a and b, chapter 10). Special Development Areas and Development Areas were merged and their extent cut between 1982 and 1984 so that only 15 per cent of the working population were now in the new combined areas (whose average unemployment was 19.6 per cent) compared to 10 per cent of the working population in the new Intermediate Areas (whose average unemployment was 15.9 per cent).

A 1987 review resulted in the abolition of both the Intermediate and Development Areas. However, these areas remain as 'designated', in order to qualify for European Community Regional Development Fund (ERDF) assistance. Now aid is **discretionary assistance** rather than an automatic grant, as it had been until 1987. Companies will only receive aid if they can prove that, without such assistance, they would not operate in the regions. There is a separate policy for Northern Ireland.

This switch in policy reflects a change in ideology from government aid to private investment as the basis of regional growth.

## Urban-based policies and targeting

Since 1976 there has been a switch to urban-based policies of regional development. New Towns represented an early form of government targeted investment, e.g. Corby (see chapter 11). Since the 1968 **Urban Programme**, there has been a gradual switch of investment to urban areas. This increased after 1976 following the **Inner Urban Areas Act**, with **Partnership Areas**, **Programme Authorities** and **Designate Districts** (see chapter 11).

The 1980 **Local Government Planning and Land Act** introduced:

1 Enterprise Zones,
2 Urban Development Corporations (UDC) (see chapter 11).

The 1981–83 **Merseyside Task Force** was a response to riots in 1981, and is now part of a Department of Environment Programme.

In 1983–84 the Urban Programme was cut back as new Enterprise Zones were established. Inner city riots in 1986 were followed by the 1987 designation of five UDCs and three mini-UDCs. In 1987 it was decided that the Enterprise Zone experiment was not to be extended in England.

## New private initiatives in regional development

Since 1979 the ideological switch from government-funded to privately financed regional development has been encouraged by the following measures:

1 The establishment of **Enterprise Boards** by local authorities, e.g. West Midlands Enterprise Board. These are now independent of local authority control.
2 The growth of **Inward Investment Organisations**, e.g. Northern Development Company.
3 The growth of **Private Venture Capital Organisations**, e.g. Northern Investors, Newcastle.
4 The various Confederation of British Industry Task Force Teams set up since 1987, e.g. in Newcastle and Birmingham.
5 **Business in the Community**, established initially (1987) in Calderdale and Blackburn.
6 **City Action Teams** (1988) in Newcastle, Nottingham and Leeds.
7 **Compacts** between enterprise agencies and local authorities, e.g. East London.
8 **City Grants** (1988), a government scheme to replace urban development and regeneration grants.
9 **Private enterprise new towns** – many proposals have been received from building companies, e.g. Foxley Wood (Hants), Bradley Stoke (Avon), Stone Bassett (Oxon) and two near Cambridge. Tillingham Hall in the Essex Green Belt and, in 1989, Foxley Wood, were rejected.

> " Private initiatives in regional development. "

## REGIONAL DEVELOPMENT IN WESTERN EUROPE

The **European Regional Development Fund** 1975 (ERDF) is the European Community's response to British pressure for a regional fund. Prior to that, the **European Coal and Steel Community** (ECSC) administered grants to convert the local economy and to retrain workers in depressed coal-mining and steel-producing regions, and the **European Investment Bank** (EI) had channelled investment into areas. The ERDF operates on the principle of **additionality**, i.e. the area must *already* qualify for national funds – hence the UK retention of its Development Area designations. Investment aid ranges from 20 per cent to 75 per cent in Northern Ireland, the Republic of Ireland, the *mezzogiorno* and West Berlin. There is a **quota system** which allocates most to the UK and Italy, but many feel that the investment is too spread out and poorly targeted.

**Integrated Development Operations** (IDO) was established in 1988 by the European Community to target assistance to specific places: Belfast, Birmingham and Naples were the first three places to receive such aid.

## CASE STUDIES OF REGIONAL DEVELOPMENT

### The United Kingdom

In the United Kingdom you should at least be aware of the effects of government policies on *one* Development Area, e.g. Central Scotland, the North-East, or South Wales. You should know *why* the area was designated and be able to use terms such as **'deindustrialisation'** and **'depressed region'** accurately, illustrating the principles with *examples* of factory closures, government schemes and their impact. You should also be able to make an attempt to 'measure' the success of a policy and to say why it was successful or failed.

In the case of South Wales you should be able to outline:

- The effect of the depression on the area: the designation of Special Areas, e.g. Trading Estate at Trefforest.

- The help given between 1945 and 1979: the aid to industry (e.g. at Ebbw Vale), new industrial estates (e.g. Bridgend and Hirwaun) and office decentralisation (e.g. motor taxation to Swansea). Be aware of new roads (M4), of rail improvements, of new government/nationalised factories (e.g. Llanwern, the New Town at Cwmbran), of reclamation of derelict land (Lower Swansea Valley), etc.

- The continuing problems facing the coalfield valleys. However, note the strength of Cardiff as a growth point on the M4 axis, the reclamation of mine areas, Japanese branch plant investment (multinational company investment), the Swansea Enterprise Zone and the Cardiff Bay UDC. Also be aware of new basic industrial investment, e.g. the proposed coal mine at Margam.

- The reasons why growth has polarised to the coastal area whilst decline continues on the coalfield. Refer to the *theories* which might explain this.

    You could adopt a similar format for most coalfield industrial areas in Great Britain, France, Belgium and West Germany (the Ruhr is a case study area for JMB 'C'), as well as the Appalachian coalfield in the USA.

### Italy

Question 2 at the end of this chapter provides a set of maps which identify the basic twofold division of Italy. Perhaps you could now attempt the question you will find there, using your revision knowledge to assist you. The one concept that we have not dealt with here which is relevant is that of **peripherality**, both in a national and a European sense (see chapter 13 for a model which uses peripherality).

## DEFINING REGIONS

Geographers talk of:

- **Formal regions.** These have *uniformity* derived from any one characteristic: geology, e.g. the Mendips; climate, e.g. the Mediterranean; land use, e.g. the Rhine gorge; etc.

- **Functional regions.** These are regions of interdependent parts, such as a **drainage basin**, e.g. the Po Basin, or a **city region**, e.g. the London region. Functional regions make the best units for physical planning, e.g. South Wales. They have therefore been widely used for planning purposes. Both Hannover and Braunschweig have had city regional planning, but were disbanded in the 1970s. The break-up of the GLC is another case where the statutory body for planning (and other functions) was disbanded. However, in the case of the GLC this was for reasons other than a dislike of planning in a functional region! The plans for Paris have long been based on a functional region.

## PHYSICAL PLANNING IN GREAT BRITAIN

The lower, smaller scale of planning is usually more concerned with the environment, whereas regional planning is concerned with the economy.

### Structure Plans

Structure Plans were introduced following the 1968 Town and Country Planning Act and are prepared by the county council. A structure plan is a broad-brush approach which brings national needs into a local context. It comprises a written statement based on a survey of the area and its needs. This provides a framework for local plans.

### Local Plans

The Local Plan applies the policies outlined in general in the Structure Plan to a district and sketches in the details of a policy. This policy is then used for development control and to co-ordinate development and change in an area. It is the duty of planners (at both levels) to seek a high degree of public participation – the involvement of people in the decision-making process.

**" Types of local plan. "**

Local Plans may take the following forms.

- **Action Area Plans**: a small area, short-term plan to achieve a specific goal, e.g. the rehabilitation of a former industrial site.
- **District Plan**: the comprehensive plan for an administrative area designed to provide the framework for both large and small-scale changes in the district.
- **Subject Plans**: a plan for a specific *use*, such as a marina development or a retailing outlet.

All planning passes through a **Development Control Process** (Fig. 12.1). If a district or county loses, on appeal, they pay the costs; likewise for any losing developer. Therefore planning applications often involve the prolonged participation of planners, politicians, developers and the public, often through **pressure groups**. These are voluntary bodies which start as interest groups, e.g. a naturalists' society, and become pressure groups when their interests are threatened. Some are permanent and are always there to influence government, e.g. the Council for the Preservation of Rural England, while others only form to oppose a particular development, e.g. a new road.

There is **bargaining** in development planning which often results in **planning gain**, e.g. a developer will provide some other facility for the public; those engaged in building a superstore may build an access road or roundabout, or even a swimming pool for the district. In all the cases of development planning where there has been conflict, it is worth asking:

- Who **gains** what, where, why and how?
- Who **loses** what, where, why and how?

This is the basis of **welfare geography**.

### Other planning controls

There are many other forms of planning control and you should be aware of them. Some have been introduced elsewhere, e.g. Green Belts (see chapter 11). There are a few other areas where controls exist (these controls are for England and Wales – similar controls exist for Scotland and Northern Ireland):

- Historic Buildings and Conservation Areas – the **1980 Local Government and Planning Act** is the present basis.
- Nature Conservation – the **1981 Wildlife and Countryside Act** is the current legal basis. Sites of Special Scientific Interest (SSSI) can be designated by the Nature Conservancy Council.
- **Heritage Coasts** – established by the Countryside Commission since 1970.
- **Areas of Outstanding Natural Beauty** (AONB) – designated since 1949 by the Countryside Commission.
- National Parks – the **1949 National Parks and Access to Countryside Act** established ten parks covering 9 per cent of England and Wales. Each park is now administered by a special board which has a planning function.

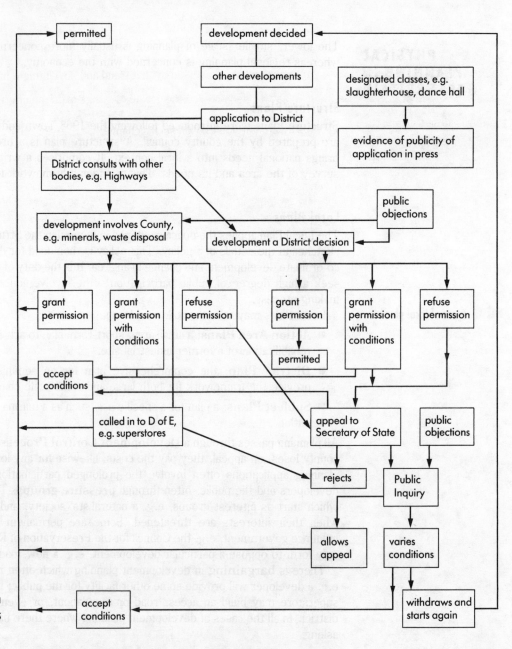

Fig. 12.1 A simplified flow diagram of the development control process

# EXAMINATION QUESTIONS

## Question 1

Study the table below which shows changes in per capita personal income for the regions of a Western country between 1900 and 1980. The level of personal income in each region is measured in terms of the national average which is indexed as 100 for each of the sample years.

| REGION | PER CAPITA PERSONAL INCOME (100 national average for each year) | | | |
|--------|------|------|------|------|
| | 1900 | 1940 | 1970 | 1980 |
| A | 135 | 127 | 108 | 105 |
| B | 143 | 132 | 114 | 106 |
| C | 107 | 112 | 105 | 103 |
| D | 98 | 81 | 94 | 97 |
| E | 51 | 77 | 90 | 93 |
| F | 50 | 49 | 74 | 79 |
| G | 61 | 64 | 85 | 95 |
| H | 140 | 87 | 90 | 94 |
| I | 163 | 135 | 111 | 113 |
| Mean deviation (%) | 37 | 28 | 12 | 9 |

Table 12.1 Per capita personal income by region (1900–80)

a)  i) Identify the nature of the trend shown by the mean deviation figures between 1900 and 1980. (1)

    ii) Explain what is revealed by that trend and how it might have come about. (5)

b)  i) Which region showed the greatest change in its ranking as between 1900 and 1980? (1)

    ii) How might you explain such a change? (5)

c)  Suggest and justify a classification of these regions based on the trends shown in the table.

    Your classification should contain at least four classes. (8)

(WJEC, June 1988)

## Question 2

Study Figs 12.2 (A–F) which show the patterns of infant mortality for 1970 and 1977, hospital bed availability, dependency ratios and youth unemployment in 1977, in the administrative regions of Italy.

a)  A Spearman's rank correlation coefficient was calculated for the data on infant mortality in 1970 and 1977. The result was r = +0.8. Using this coefficient and the maps, describe the relationship between the patterns of the two variables. (5)

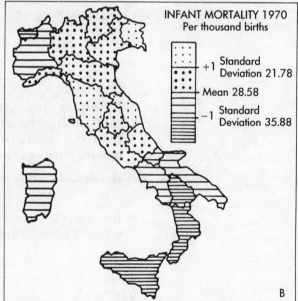

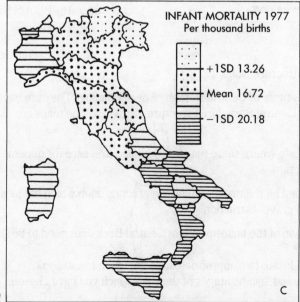

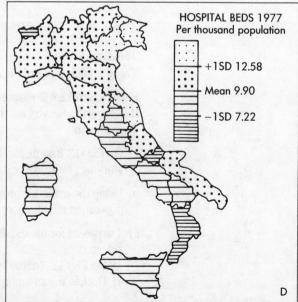

Fig. 12.2

Fig. 12.2 (continued)

b) A Spearman's rank correlation coefficient was calculated for the data on infant mortality in 1977 and the number of hospital beds per thousand population. The result was r = +0.67. Using this coefficient and the maps, describe the relationship between the patterns of the two variables. (5)

c) By reference to diagrams E and F in Fig. 12.2, outline the geographical patterns of dependency and youth unemployment. (5)

d) How does the information shown on the maps relate to your understanding of the theory of national cores and peripheries? (10)
(AEB)

### Question 3

Assess the effectiveness of regional development policies within ONE *named* country.
(London 16–19, A/S Sample)

### Question 4

This question is based on a Structure plan.

Figs 12.3 to 12.6 illustrate features of Central Berkshire near Reading. They are based on the Report of Survey for the Central Berkshire Structure plan 1976. The maps are diagrammatic.

a) On Fig. 12.3 name and locate where there may be continued pressure for mineral workings for gravel extraction. (4)

b) Using the evidence provided on the maps, outline the grounds upon which the local population might object to gravel extraction. (9)

c) Two areas for the expansion of the built-up area of Central Berkshire need to be selected.
  i) On Fig. 12.3 name and locate two appropriate areas for such expansion. (4)
  ii) Outline the advantages and disadvantages of the sites which you have chosen. (8)
(AEB)

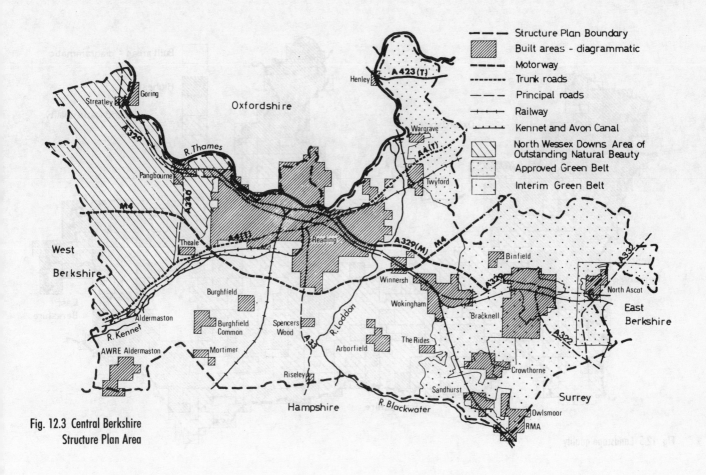

Fig. 12.3 Central Berkshire
Structure Plan Area

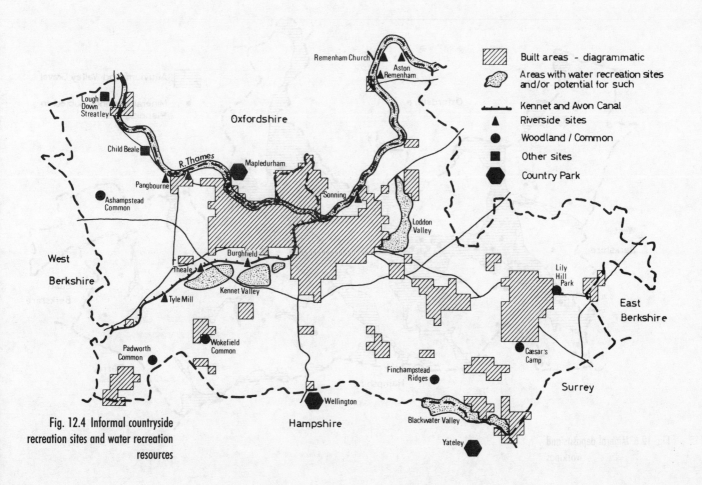

Fig. 12.4 Informal countryside
recreation sites and water recreation
resources

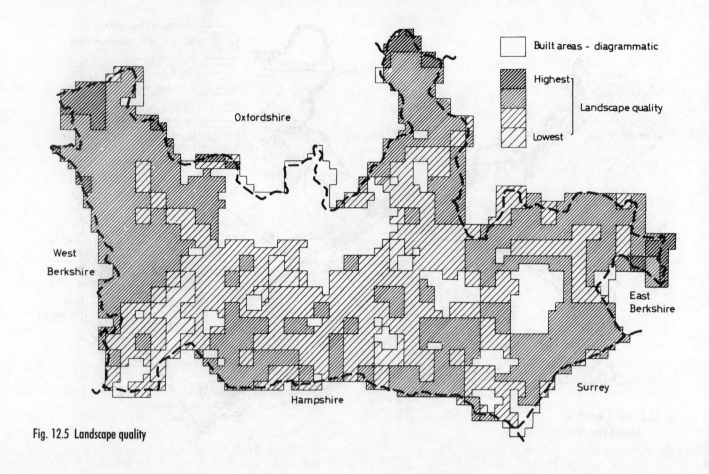

Fig. 12.5 Landscape quality

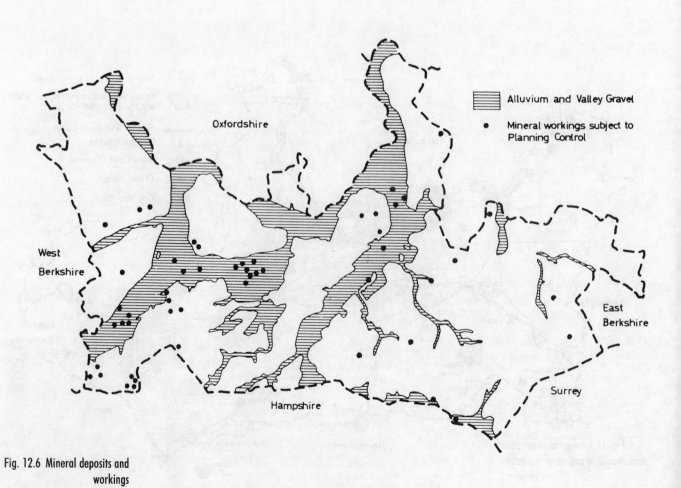

Fig. 12.6 Mineral deposits and
workings

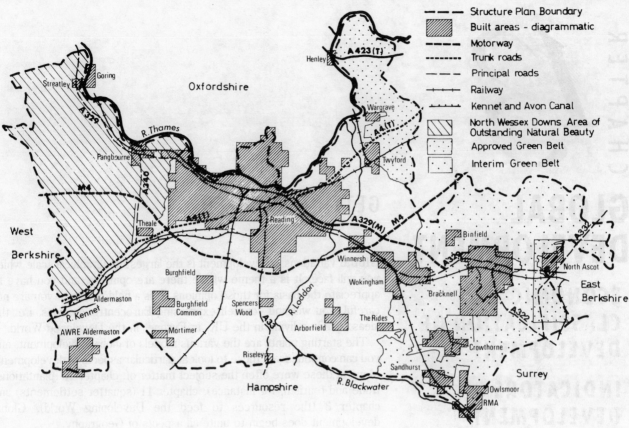

**Fig. 12.7 Central Berkshire Structure Plan Area**

Legend:
- — · — · — Structure Plan Boundary
- Built areas - diagrammatic
- — — — Motorway
- ----------- Trunk roads
- — — Principal roads
- ┼─┼─┼ Railway
- ┼─┼─┼ Kennet and Avon Canal
- North Wessex Downs Area of Outstanding Natural Beauty
- Approved Green Belt
- Interim Green Belt

# OUTLINE ANSWER

## Question 1

a) Mean Deviation measures the dispersion of a data set around the arithmetic mean. The figures are getting less dispersed. There is less regional divergence; this could come about either as a result of regional policy or as a product of market forces.

b) H changed most. It could be the loss of a resource, thereby affecting the economic base. The closure of a major employer during the depression years is another possibility.

c) A classification will need four types:
1 Those above the national average;
2 Those below the national average;
3 The zone of rapid decline;
4 (1) and (2) can be subdivided to exclude C and D and place these in a separate category of little change relative to the mean.

Justification should contain some reference to (1) Core and periphery, (2) the role of regional planning, and (3) the ideas of cumulative causation maintaining the relative rankings of all regions. You could refer to example regions that you know, even though the question does not demand it.

# CHAPTER 13

# GLOBAL DEVELOPMENT

**ECONOMIC CLASSIFICATION OF DEVELOPMENT**

**INDICATORS OF DEVELOPMENT**

**OBSTACLES TO DEVELOPMENT**

**STRATEGIES TO OVERCOME OBSTACLES**

**DEVELOPMENT STRATEGIES**

## GETTING STARTED

**Global** variations in development is the largest geographical scale which you will face. It is a theme where there are opinions, and you have to appreciate the basis for those opinions. It is a field where, if you are not careful, you will find yourself expressing Eurocentric opinions, i.e. the ideas based on living in the UK, in Europe, in the Developed World.

The starting points are the various models of global development, and you can continue from these to look at particular aspects of development. Many of these were often the subject matter of: chapter 10 (plantations, trade and tourism, for instance), chapter 11 (squatter settlements) and chapter 3 (the resources to feed the Developing World). Global development does begin to unite all aspects of Geography.

Development involves progress in a number of directions:

1 Economic development towards a stronger economic base;
2 Technological development, thereby increasing the ability of machines to benefit people's lives;
3 Development in the welfare of the population – health, well-being;
4 The modernisation of society to fit into the twentieth century.

The **Developed World** or **First World** is sometimes referred to as 'the North', taking the term from the Brandt Report (1980) *North–South: A Programme for Survival*. These are the countries of the advanced world whose wealth had developed over a long time and which are normally capitalist countries. The countries of the North have the private ownership of capital and an economy driven mainly by private profit.

The **Developing World** is often called the **Third World** or the 'The South', and refers to the poor, underdeveloped countries in Africa, Asia and Latin America. These countries are still in the early stages of economic, technological and social development and include the NICs (chapter 10). Often they are called LDC's – Less Developed Countries.

The **Second World** is a term often used to distinguish the socialist countries from the capitalist ones in the more developed North. Not all socialist countries are part of the Second World, e.g. Tanzania and Vietnam.

Some people refer to the **Fourth World** of the LLDC, least developed countries, those whose economic prospects lag so far behind that they are caught in a vicious circle of increasing poverty (see below). Even these are being subdivided to include a **Fifth World**, where Malthusian checks are beginning to bite (see chapter 4), e.g. Chad and Ethiopia.

# ESSENTIAL PRINCIPLES

**ECONOMIC CLASSIFICATION OF DEVELOPMENT**

The World Bank (Fig. 13.1) has its own classification, which is based on economic factors.

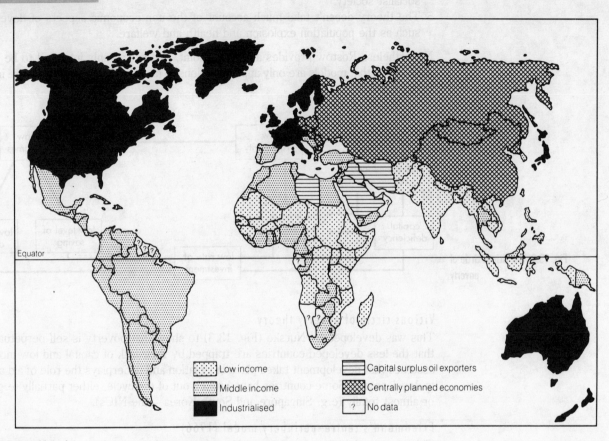

Low income

Middle income

Industrialised

Capital surplus oil exporters

Centrally planned economies

? No data

Equator

**Fig. 13.1 The World Bank groupings of countries, 1981**

## Rostow's stages of economic growth model

This model (Fig. 13.2) suggests that a country passes through five stages on its way to economic maturity.

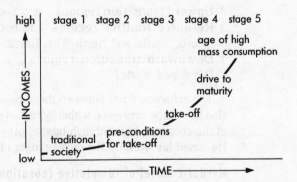

high    stage 1  stage 2  stage 3  stage 4  stage 5

INCOMES

low

age of high mass consumption

drive to maturity

take-off

pre-conditions for take-off

traditional society

TIME

**Fig. 13.2 The stages of economic growth model (after Rostow)**

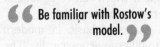

**Be familiar with Rostow's model.**

- A **traditional society** dominated by subsistence agriculture.
- The **pre-conditions for take-off**, where trade expands and new inventions/technology are introduced.
- The key stage – **take-off**, when modern methods have replaced the traditional ones with increased investment, so that growth is self-sustaining.
- The **drive to maturity** during which a more complex pattern emerges in an urban society.
- Finally the **age of mass production and mass consumption**.

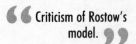

> Criticism of Rostow's
> model.

The model has been criticised in a number of ways:

1 It is based on Europe and North America over the past centuries, i.e. it is **Eurocentric**.
2 It assumes that all states pass through this sequence.
3 It over-emphasises the role of capital in development when there are countries that attempt to develop without the capitalist ethic, e.g. Tanzania.
4 The stages do not take account of resource transfers in the form of aid which have geopolitical motives: i.e. either to stop the advance of communism or to reinforce a socialist society.
5 The theory doesn't take much account of the non-economic aspects of development, such as the population explosion and health and welfare.

Nevertheless Rostow provides a starting point, and it was only intended to be a model. Remember that models are only approximations. The exceptions might be more important than the rules.

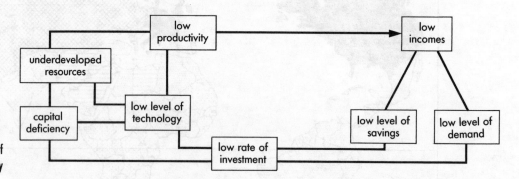

Fig. 13.3 The vicious circle of poverty

### Vicious circle of poverty theory

This was developed by Nurske (Fig. 13.3) to show that poverty is self-perpetuating and that the less developed countries are trapped by their lack of capital and low incomes. It assumes that development takes place in isolation and underplays the role of aid and loans in development. Some countries have broken out of the cycle, either partially (e.g. China) or almost totally (e.g. Singapore and South Korea – the NICs).

### Friedmann's centre–periphery model (1966)

This argues that development is dependent on the spread of the means to advance economically from the developed core regions or countries to the less developed periphery.

The model divides the world into four:

1 **Core regions** – Western Europe, North America, Japan;
2 **Upward transition regions** – e.g. Southeast Brazil, South Korea, Israel and Kuwait;
3 **Resource frontier regions** – in both the developed and developing worlds, e.g. Siberia, Alaska and North-West Brazil.
4 **Downward transition regions** – e.g. Chad, Ethiopia and declining areas in the Developed World.

The economic flows between these regions are shown on Fig. 13.4. Friedmann's model also has a time dimension within individual states (Fig. 13.5) which shows the development of the economic links through time to provide an integrated, developed economic system. He based his work on the development of Venezuela in the twentieth century.

### Myrdal's model of cumulative causation (1975)

This is a less optimistic view of the processes of development which can be applied at both the global and national scales. (This model is also covered in chapter 12.)

The theme of similar processes operating at different scales is basic in modern geography. Myrdal believed that regions and countries tend to diverge economically. The spread effects of benefits from the Developed World would be exceeded by the **backwash effects** which pull resources and people to the core region. The model is circular and cumulative because the pull of capital, resources and people to the core causes a downward spiral of decline and increasing deprivation in the periphery LDCs, and a corresponding upward spiral in the Developing World. Governments and charities try to reduce the backwash effect.

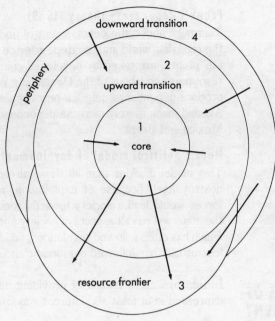

Fig. 13.4 Friedmann's centre –
periphery model

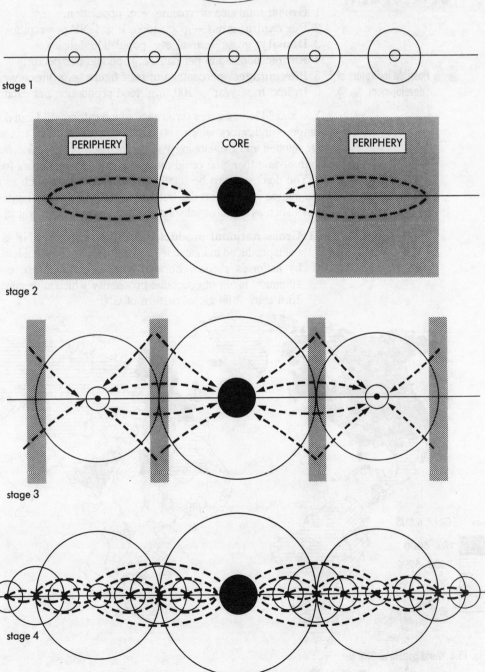

Fig. 13.5 Friedmann's development
model

### Frank's dependency theory (1969)

This talks about 'the development of underdevelopment' in terms of the **domination** of the capitalist world and the **dependence** of the Third World. Surplus value of production, i.e. profits, are taken by developed countries and the transnational companies from the resources and labour of the Developing World. Frank maintains that capitalism will always produce dependence unless a new economic and social order can break that dependence. Aid and loans merely increase dependency – see the debt problems of countries such as Mexico and Brazil.

### Marx's political model of development

This model is older than all those already mentioned. Marx said that capitalism would destroy itself because of exploitation, inequality and uncontrolled competition. These forces would lead a society towards socialism and eventually towards communism when the state will run all aspects of society and its economy for the benefit of all. This is a path which has often followed revolutions, which then jump the capitalist stage, e.g. in China, or less usually has followed democratic choice, e.g. Tanzania.

**INDICATORS OF DEVELOPMENT**

**Indicators** are a means of providing us with measures of development. They can be expressed in at least six different ways:

1 **Gross**: total size or volume, e.g. population.
2 **Per capita**: gross ÷ population, e.g. GNP per capita.
3 **Density**: gross ÷ area, e.g. population density.
4 **Ratios**: population per data, e.g. people per doctor.
5 **Percentages**: percentage of total figure, e.g. literacy rate.
6 **Index**: base year = 100, e.g. food production per capita 1980 = 100.

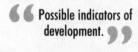

*Possible indicators of development.*

You should have figures on a range of indicators. All I can do here is give you a very broad range of indicators which you might like to use. Try to remember an example from the Developed and Developing Worlds for each of these indicators. If you are really good at memorising then you could memorise data for examples from the First to Fifth Worlds!

You don't have to be totally accurate with data such as indicators. The right order of magnitude will gain credit, but vague statements like 'bigger' or 'smaller' or sensationalist ones such as 'vast' or 'absolutely awful' gain little credit in examination answers.

1 **Gross national product (GNP) per capita**: GNP is the total value of all that has been produced in industries and services by the resident population or transferred to it by nationals abroad. Foreign worker earnings are excluded. GNP per capita is a summary index of economic prosperity which is usually expressed in US dollars. Fig. 13.6 shows the global pattern of GNP.

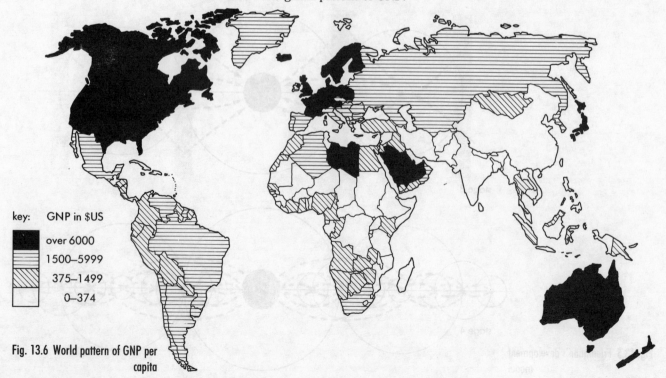

key: GNP in $US

- over 6000
- 1500–5999
- 375–1499
- 0–374

**Fig. 13.6** World pattern of GNP per capita

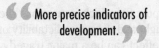

> More precise indicators of development.

2 **Gross domestic product (GDP):** This measures the total value excluding that of foreign workers' remittances and is not so suitable for measuring national income or wealth. Nevertheless it is used.

3 **Percentage of GDP from economic sectors** measures the reliance of the economy on the various sectors and can be related to the various stages of growth.

4 **The shares of exports from primary products** is a further measure of the stage a country has reached. For instance, it indicates whether the country is likely to have the capital to develop the processing of primary products.

5 **Labour force data** for each sector of the economy also indicates the degree of economic development, as in (3) above.

6 **The value of exports per capita** is a measure of how much wealth is coming into the country from trade.

7 **Energy consumption per capita** is normally expressed as kilogram of coal equivalent and doesn't include items such as firewood. It is a measure of the use of commercial energy, which is itself an indicator of industrial development.

8 **Cement production** (kilogram per capita) measures the investment in new buildings and in infrastructure such as roads.

9 **Percentage of population living in urban areas** tells us about the degree of advancement in both economic and social terms. It is a key aspect of Rostow's drive to maturity and of the development of the periphery.

10 **Cars per thousand people** is a very Eurocentric view of development. However, one cannot deny that it does indicate the wealth to spare after essentials.

11 **Telephones per thousand people** is an indicator of the communications development of a society. Those entering the later stages have more transactions and more technology – using telephones. It is also a measure of personal wealth.

12 **TVs per thousand people**

13 **Daily newspapers per thousand people**
Both of these are obvious. Are they good indicators? They do indicate broad levels of consumption and consumer awareness.

14 **Infant mortality per thousand population** is an indicator of the investment in health care.

15 **Life expectancy** (chapter 3), again measures the care system of a society.

16 **Birth rate per thousand** is a further demographic indicator of social development.

17 So too is **death rate per thousand**, and

18 **Natural increase of population** (percentage).

19 The **population under the age of 15** is a further indicator of social development. It is linked to the demographic concept of **dependence** outlined in chapter 3.
(If these terms are vague, why not go back to chapter 3. Once again, Geography is training you to interrelate your knowledge.)

20 **Food intake in calories per capita** is a further measure of health and of the ability of a society to feed itself.

21 The **index of food production** gives a time or temporal dimension to food output. If it is rising, either technology or other resources in the economy are providing more food. If it is declining, then the economy is probably declining and Malthusian checks (chapter 4) might be imminent.

22 **Percentage of population with access to safe water**,

23 **Population per doctor**, and

24 **Population per hospital bed**, all measure investment in welfare, which comes from having a surplus in an economy.

25 **Adult literacy** (percentage of population) is a measure of educational development.

## OBSTACLES TO DEVELOPMENT

It is always probable that there is more than one reason for a lack of development. Obstacles are short-term, e.g. Sahel drought, and long-term, e.g. lack of capital. The obstacles can come from within the country, such as political unrest in Mozambique, or from outside the country, as in the case of low prices for tea, which is beyond the control of Sri Lanka.

### ENVIRONMENTAL OBSTACLES

These are partly influenced by population pressure (chapter 4), increased demands for fuel and firewood (chapter 5) and extensions of farmland into marginal areas (chapter 10). Such an extension increases the probabilities of soil erosion (chapter 9), which in turn has the likely outcome of a decreased productivity per hectare. Drought and the unpredictability of rainfall is a further cause for concern. In the long term, the effect can be a major tragedy, as has occurred in the Sahel.

> **❝ Different obstacles to development. ❞**

**Desertification** (chapters 6, 9 and 10) is also a major problem. In contrast to drought there is **flood**, caused, for example, by the monsoon which has catastrophic short-term effects in terms of crop and livestock losses. Similarly hurricanes and typhoons have an adverse effect on the fragile economic system in their path, destroying crops (e.g. the banana crop in the West Indies), flooding coastal areas (e.g. in Bangladesh) and destroying fishing vessels (e.g. South China Sea).

## COLONIALISM

Colonialism is the rule over peoples of the Developing World by the European powers, especially during the late nineteenth century. This did affect the development of much of Africa and South Asia in particular (Fig. 13.7). The colonies were dependent to a greater or lesser degree as the colonial powers introduced agricultural systems which annexed their mineral resources, converted their farmlands or opened up new lands for plantation (chapter 10). They used the native people as slave labour (Barbados) or as cheap labour, even importing cheap labour in the case of Malaysia to work in the tin mines and rubber plantations. Peoples were moved off their lands so that white settlers could farm the best lands, e.g. White Highlands of Kenya. Cash crops and plantations led to monoculture, which was not a problem for a colony but, on independence, such dependence on a single crop was economically dangerous because of fluctuations in the world demand and price paid for the crop. Many former colonies were left with too narrow an economic base and too little diversity of export earnings.

> **❝ Excessive concentration can be dangerous. ❞**

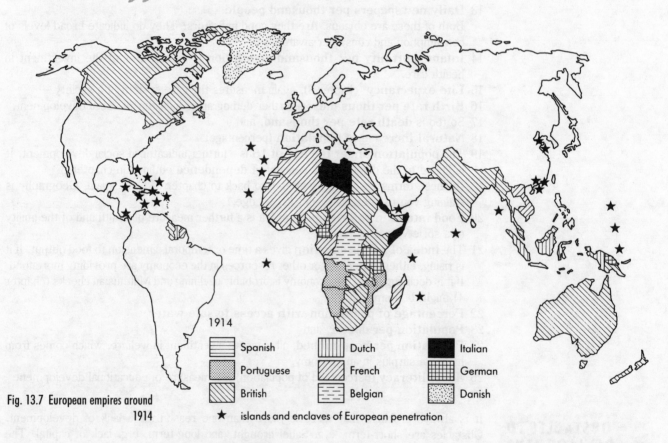

Fig. 13.7 European empires around 1914

1914

| | | | | | |
|---|---|---|---|---|---|
| ▤ Spanish | ▥ Dutch | ■ Italian |
| ▦ Portuguese | ▨ French | ▦ German |
| ▧ British | ▤ Belgian | ▥ Danish |

★ islands and enclaves of European penetration

## NEO-COLONIALISM

Despite independence neo-colonialism is now said to exist; it describes the tendency for the countries of the Developed World to maintain their influence over their former colonies because the previous patterns of trade have continued. The products of the new states are frequently bought by the multinational companies based in the Developed World, e.g. Lonrho. Alternatively the loans made to the former colonies often originate from the banks of the Developed World, e.g. Barclays.

## SHORTAGES OF CAPITAL

This is a root cause of the problem, according to Rostow. There is a resulting trade problem because the benefits of specialised production are largely focused in the Developed World, with its access to capital investment and large-scale enterprise. The Developing World is unable to specialise enough to equalise the balance of trade, partly because it lacks the capital to invest in major projects.

## TRADE BARRIERS

These restrict access to markets in order to protect the Developed World (see chapter 10 for a note on trade blocs such as the European Community). Trade agreements and protocols (such as that for sugar) only partly alleviate the adverse trade pattern caused by tariff barriers on cane sugar from Barbados and Mauritius. These barriers are designed to protect the beet-sugar growers of Western Europe.

## GOVERNMENTS

Governments are often the unwitting obstacles to development. The policies followed by many developing countries focus development in one area, as occurred in Venezuela and South-East Brazil (see Friedmann, earlier in this chapter, for the theory based on this). The focus upon growth poles (see chapter 12) and growth regions, to the detriment of the peripheral regions, causes a range of economic and social problems in both the core and the periphery. Over-rapid urbanisation (chapter 11) with shanty towns, urbanisation without industrialisation, and rural depopulation, are just a few of these problems. Government expenditure is often unduly concentrated on itself. The Central African Republic, defending itself against overthrow and revolution, is perhaps one example, and Ethiopia, for similar reasons, another. This diverts potential government investment away from the economy and can deter investment and aid from overseas. Various forms of unrest, ranging from blatant racialism in South Africa (which has created the Homelands as islands of poverty) to inter-community violence in Sri Lanka, also deter investment. Other political conflicts have also resulted in lack of investment in e.g. Vietnam. Countries offering loans prefer the sound investment of an NIC, especially in those states whose political system is 'in favour' with the lenders' government.

## TRANS-NATIONAL CORPORATIONS

These have a major influence on development and some see them as the major agents of neo-colonialism. Countries try to attract these companies with free port facilities (e.g. Singapore) and other trading concessions. Multinational companies can switch productive capacity between countries, often to the detriment of their host economies. Very often the turnover of these companies exceeds the total value of the trade of the developing country, and so the company has the greater economic muscle.

## DEMOGRAPHIC OBSTACLES

You should recall what Malthus hypothesised, which is used by the neo-Malthusians today (such as Erlich) to forecast an increased problem of overpopulation. However, others ask whether starvation is rather caused by not enough food being in the right place at the right time. When famine hits the headlines, the fact that the food aid is elsewhere is often mentioned. (See chapters 3 and 4.)

## STRATEGIES TO OVERCOME OBSTACLES

## TECHNOLOGICAL TRANSFER

This is the movement or sale of technologies for industrial and agricultural development to the developing countries. This is the strategy of the transnational companies who make, for example, Grundig TVs in Malaysia. The countries can insist that they share in the adoption of the technologies in this way. There is then an increased benefit to the developing nations. For example, both Singapore and Malaysia insist on some local equity in companies. In the tourist trade, the Malaysian government has a share in many hotel/resort developments and MAS, the national airline, is now targeting package holiday companies.

## INTERMEDIATE TECHNOLOGY

This is the concept that both capital- and labour-intensive technologies are inappropriate in the Developing World. Schumacher argued that intermediate technology would bridge the gap between the old, more primitive technologies, and the advanced technologies. Such schemes now abound in many areas, such as the simple irrigation systems and rope wells funded by charities such as Bandaid. The promotion of local crafts, such as coconut-shell carving in St Lucia and batik cloth designs in Malaysia, are other schemes which cost little for substantial development gains.

## INDUSTRIAL STRATEGIES

This is the solution of the NICs and of countries like India and China where industrialisation has sustained economic growth, despite the rapid growth in population. Industrialisation is capital intensive, and is often based on credit from banks for the high fixed costs of quality machinery. Industrialisation therefore depends on substantial government assistance and the involvement of foreign governments and the multinational corporations. Skilled labour is essential and therefore there must be a good education system. Successful industrialisation then depends on the willingness of countries to *trade* in the new industrial products. This has normally already been agreed, because of the foreign involvement in the first place. Some even have enterprise zones to attract investors, e.g. in China on the Hong Kong border.

## IMPORT SUBSTITUTION

This is the replacement of costly imports by locally made produce. It enables diversification and the acquisition of skills, but again some profit is lost if no local equity is used.

## AGRICULTURAL DEVELOPMENT

(See the Green Revolution in chapter 10.) Mechanisation of agriculture is often extensive, even with intermediate technologies. The danger of mechanisation is underemployment in the villages, which is a push factor in urban migration (chapter 3).

## RURAL DEVELOPMENT PROGRAMMES

These can be targeted at various types of development, but they need finance either from export earnings or from aid. There is a fund to help raise food production by reforming the patterns of divided holdings which inhibit the use of the simplest mechanical aids to cultivation. The provision of roads, water, electricity and waste disposal is a simple form of infrastructure development. Land reform programmes to remove absentee land-owners of vast estates have been attempted in some countries, e.g. Peru. In Cuba, the land reform was more radical and involved state farms and co-operatives (see chapter 10).

## TRADE

This is the corollary to many of the schemes which have been outlined above. Without it there is little cash for development. Protectionist policies of trade blocs must be overcome. Cartels, the most famous of which is OPEC (the Organisation of Petroleum Exporting Countries) have often been able to extract more money for their products, either by raising prices (1973) or by restricting output and so forcing up prices.

## AID

This is often seen as the alternative to trade. It can be multilateral (e.g. from the European Community) or bilateral (e.g. from one country to another). UK aid is 40:60 respectively.

The UN has recommended that aid should comprise at least 0.7 per cent of the GNP of all the developed and oil-rich countries. Aid is often politically motivated or even neo-colonial in its distribution. Aid is seen as a short-term strategy, otherwise peoples can become **aid dependent**. Aid can be 'tied', i.e. to buy the industrial products of the donor, and as such it may not benefit the recipient that much. Food aid can alter tastes and diets so that people begin to look to imported, rather than domestic, grains. The answer may well be untied aid, but that has to be monitored.

## TOURISM

(See chapter 10.)

### DEVELOPMENT STRATEGIES

- The Tanzanian Rural Development **Ujamaa** is a socialist strategy based on co-operative villages together with government-run industrial corporations and the service sector.
- Five Year Plans – India – the inspiration was Russian planning with the initial development of heavy industry (Damodur valley) together with an attempt to alleviate poverty.
- The NICs (see chapter 10).
- The Chinese Communist strategy which has removed social inequalities and improved agricultural output through a series of initiatives, such as 'The Great Leap Forward' and 'The Cultural Revolution'. More recently the 'Special Economic Zones' e.g. Shenzhen and Zuhai are development strategies.

# EXAMINATION QUESTIONS

### Question 1
By reference to selected countries, examine the relative roles of trade and aid in reducing international inequalities.

### Question 2
Study the table below which lists some socio-economic indicators for selected less developed countries in 1978.

| Country | A | B | C | D | E | F |
|---------|------|-----|------|----|----|------|
| 1 Upper Volta | 160 | 1.6 | 1880 | 42 | 83 | 30 |
| 2 India | 180 | 2.0 | 2020 | 51 | 74 | 180 |
| 3 Zaire | 210 | 2.7 | 2270 | 46 | 76 | 70 |
| 4 Tanzania | 230 | 3.0 | 2060 | 51 | 83 | 70 |
| 5 China | 230 | 1.6 | 2470 | 70 | 62 | 310 |
| 6 Kenya | 330 | 3.3 | 2030 | 53 | 79 | 140 |
| 7 Indonesia | 360 | 1.8 | 2270 | 47 | 60 | 250 |
| 8 Egypt | 390 | 2.2 | 2750 | 54 | 51 | 460 |
| 9 Zambia | 480 | 3.0 | 2000 | 48 | 68 | 470 |
| 10 Bolivia | 510 | 2.6 | 1970 | 52 | 51 | 370 |
| 11 Nigeria | 560 | 2.5 | 1960 | 48 | 56 | 110 |
| 12 Morocco | 670 | 2.9 | 2530 | 55 | 53 | 290 |
| 13 Peru | 740 | 2.7 | 2270 | 56 | 39 | 650 |
| 14 Malaysia | 1090 | 2.7 | 2610 | 67 | 50 | 720 |
| 15 Mexico | 1290 | 3.3 | 2650 | 65 | 39 | 1380 |

A:   Gross national product per head (US dollars)
B:   Population growth rate (percentage)
C:   Daily food supply per head (calories)
D:   Life expectancy at birth (years)
E:   Labour force employed in agriculture (percentage)
F:   Energy consumption per head (kilograms of coal equivalent)

Table 13.1

a) Define 'less developed country'. *(1)*

b) Suggest factors which might account for the variations in energy consumption. *(3)*

c) With specific reference to countries or regions you have studied, describe and comment on the
ways in which government policies influence:
  i) Energy consumption. *(4)*
  ii) The numbers employed in agriculture. *(4)*

(AEB, 1989)

## Question 3

a) Define plantation agriculture and bush fallowing. *(2)*

b) Outline three factors that might give rise to plantation agriculture in developing
countries. *(4)*

c) Significant changes have taken place in the ownership and management of plantations in
developing countries since they received independence.
  i) Briefly describe the changes that have taken place. *(3)*
  ii) How do the changes reflect the power structure in the countries? *(3)*

(AEB, 1989)

## Question 4

Table 13.2 gives data on agricultural production for a selection of African countries in the
1980s. A series of Spearman rank tests has been carried out on the columns of data and the
results are shown in Table 13.3. Fig. 13.8 shows the significance levels of the Spearman
rank coefficient; there were 14 degrees of freedom.

a)  i) Which TWO columns of data show the greatest degree of rank correlation? *(2)*
   ii) What do the significance levels tell us about those results showing the greatest
       degree of rank correlation? *(4)*
   iii) Explain why the TWO variables showing the most significant correlation would be
       expected to be significantly related. *(6)*

b) Using the data in Table 13.2, suggest which countries might be most overpopulated.
Justify your choice. *(8)*

c) What other data would be needed to support the choice you have made in (b)? *(5)*

(AEB, November 1988)

| COUNTRY | 1<br>Population density (persons per km²) | 2<br>Index of agricultural production 1982–4 (1974–76 = 100) | 3<br>Index of agricultural production per agricultural worker 1982–84 (1974–76 = 100) | 4<br>Calories domestically produced as percentage of total supply | 5<br>Crop yield for cereals 1982–84 (kilograms per hectare) | 6<br>Fertilisers used 1981–3 (kilograms of plant nutrient per hectare of arable and permanent cropland) | 7<br>Tractors 1981–83 (number per thousand hectares of arable and permanent cropland) |
|---|---|---|---|---|---|---|---|
| Benin | 35.6 | 97 | 112 | 110 | 678 | 2 | 0.1 |
| Burkina Faso | 25.3 | 96 | 106 | 95 | 521 | 4 | 0 |
| Chad | 25.3 | 96 | 101 | 100 | 539 | 2 | 0.1 |
| Egypt | 46.7 | 91 | 102 | 79 | 4327 | 311 | 0.1 |
| Ethiopia | 29.8 | 102 | 112 | 97 | 1210 | 3 | 0.3 |
| Ivory Coast | 30.4 | 109 | 113 | 106 | 904 | 10 | 0.8 |
| Kenya | 35.4 | 85 | 100 | 97 | 1472 | 32 | 2.8 |
| Malawi | 59.2 | 100 | 114 | 104 | 1164 | 14 | 0.5 |
| Mali | 6.5 | 107 | 116 | 96 | 742 | 5 | 0.4 |
| Nigeria | 103.1 | 96 | 123 | 93 | 372 | 7 | 0.3 |
| South Africa | 26.5 | 83 | 97 | 16 | 1178 | 87 | 13.3 |
| Sudan | 8.6 | 85 | 106 | 100 | 470 | 6 | 0.9 |
| Tanzania | 23.8 | 96 | 98 | 98 | 964 | 5 | 3.6 |
| Zambia | 8.9 | 73 | 86 | 89 | 1637 | 19 | 0.9 |
| Zimbabwe | 22.4 | 69 | 87 | 114 | 936 | 60 | 7.5 |

**Table 13.2** Selected indicators

*Source: World Resources 1986* (Basic Books, 1986.

| Columns 1 and 2 | 0.18 |
| Columns 1 and 3 | 0.4 |
| Columns 2 and 3 | 0.81 |
| Columns 3 and 5 | −0.5 |
| Columns 3 and 7 | −0.6 |
| Columns 4 and 5 | −0.18 |
| Columns 5 and 6 | 0.6 |
| Columns 5 and 7 | 0.6 |
| Columns 6 and 7 | 0.84 |

Table 13.3 Spearman rank correlation coefficient results

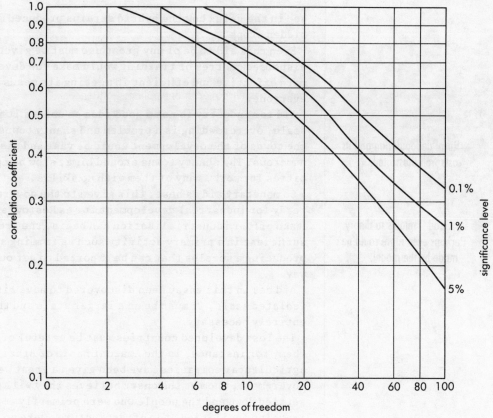

Fig. 13.8 Significance levels of the Spearman rank coefficient

## Question 5

Why is Rostow's model of economic growth inappropriate to many countries in the Developing World today?

*(25)*

(AEB, 1987)

## Question 6

Assess the contribution that an improvement in transport may make to economic development in tropical areas.

(London)

## STUDENT ANSWER WITH EXAMINER'S COMMENTS
### Question 1

Aid to the less developed countries can take many forms, money, food, equipment for farming, or industrialisation. In any case aid must be necessary to that area, for by giving direct aid we are making the less developed countries dependent on the developed.

Giving aid to one area can cause migration in from other nearby and even long distant locations. The food aid of Ethiopia, brought thousands treking to the few areas of medical help and food supplies. In such conditions seen in such a situation, there are too many people who need help and too few facilities in the little camps. Overcrowding in such conditions, brings a bitterness, and helplessness. Thus such aid centres for medical and food supplies must be numerous and dispersed, also their affect is of little help to anyone, they can not help so many people.

After the situation has stabilised in Ethiopia, we must help them to earn a living once more, they must become self sufficient, but the primary force for such agricultural developments must come from the developed world.

This aid can take the form of hybrid strains of fast growing foods, which take little nutrients from the soil, fertilisers may be introduced to the area and up-to-date farming equipment must be used. This aid can be given in one area for example in India farmers came to the development centres and were taught how to use the aid given. They could

> **Good; makes the distinction between the need for short-term crisis relief, and for longer-term structural solutions.**

> **Good points.**

see in the fields how the hybrid strains produced more than their disease ridden crops.

Even this aid to be of any great use must be given on a large scale and dispersed centres of training would make the development program much more effective and efficient in meeting its means – enough food for everyone.

Aid for industry is not a good idea as unless it is done on a large scale, overcrowding is a problem and shanty towns develop. Migration to the towns of the development would be vast and disease would be numerous. The shanty towns around Lima, show how people come to the cities for work, many of them with no skills.

So monetary aid is best, this given to the government when the area is ready for industrial development for as Rostow's model shows, before "take off" of industrialisation can begin, the area must firstly be self sufficient in a primary activity such as farming and secondly be producing a surplus that can be exported to get our financial help that way.

Aid has in this essay been discovered to have little use on a small or isolated scale, it must be on a larger scale and then only if it is entirely necessary.

The less developed countries must be careful of how that aid comes to them, for instance, in the Amazoni Basin of Brazil, foreign ranch and agricultural companies have been given a great welcome by the Government, whom at the onset believes they will gain. The foreigners settled down and the people who were primarily employed to build ranching facilities and infrastructure, were no longer necessary. It soon became clear than one man could look after a small ranch on his own, the ranches developed and the Brazilian government waited patiently for a financial gain. Any gain however it was soon found, was going abroad to be put into home projects of the foreigners. The governments were left with, if anything, a financial loss after producing the infrastructure for such developments.

It has been stated recently that the less developed countries must not search for aid from abroad in any shape or form, it only brings a financial loss to the developing country by means of paying off their debts, or having their resources exploited by foreigners. The third world has to face the fact that the first world has almost "shut the door on it", The developed world only looks towards its own gain and not in the help of others.

Thus it must be concluded that really if aid can be avoided it must not be taken, and if it is entirely necessary it must be employed on a geographically dispersed development, rather than the development of one region. As the development region brings migration from other areas and a poorer environment and standard of living in the long run. Aid is sometimes a necessary evil, but it is in some cases only giving from the developed to the less developed countries.

> " Rural depopulation and its costs well illustrated. "

> " Good – relates to theory. Perhaps even more could be made by the model. "

> " A bit long-winded, but an example of aid having little 'trickle down' effect. Actually it's inward investment rather than aid! "

> " A bit confused! Little justification here. "

> " Good point about dispersed development, but little else considered here! "

> " The answer has rather neglected trade and looked almost exclusively at aid. Some useful examples given. Too little use has been made of theory, or of relevant examples of theory, for this answer to receive a high mark. "

# THE PROJECT PAPER

## GETTING STARTED

A field project has become an essential component of both A and A/S-level Geography papers in recent years and it will not be long before all Geography examinations will require the submission of an individual piece of work. How this work is assessed and how much it contributes to the final grade does vary, as you can see in Table 14.1. Nevertheless, it is an important part of your studies and careful preparation can help you to obtain a good grade. It is also the earliest of your assessments, in that you will begin it during or towards the end of your first year and hand it in about Easter before your written papers.

| A-level Board | Whether project paper | Alternative to | Maximum No. of words | % of final mark | Board- or teacher-marked | Board moderation | Outlines submitted & date | Oral exam |
|---|---|---|---|---|---|---|---|---|
| AEB | √ | – | 5000 | 20 | B* | – | Y | X |
| Cambridge | √ | P4 | | 28 | B | – | √1/11 | √ |
| London (210) | V | – | 5000 | E | T | √ | – | – |
| London (16–19) | √ | – | 3500 | 11 | T | √ | √15/5 | – |
| JMB 'B' | | Both syllabuses have a teacher marked, board moderated alternative to the practical paper. | | | | | | |
| JMB 'C' | | It could involve practical field projects. | | | | | | |
| Northern Ireland (NISEC) | √ | – | | 15 | T | √ | – | – |
| Oxford | √ | – | 4000 | 28 | B | – | √1/11 | – |
| Oxford & Cambridge | √ | P6† | 1000 | 25 | B | – | √1/10 | √ |
| Scottish Higher | √ | – | 1000 | 25 | B | – | – | – |
| WJEC | √ | | 4000 | 15 | T | √ | – | – |
| **A/S-level** | | | | | | | | |
| AEB | √ | – | 3000 | 30 | T | √ | √Y | – |
| COSSEC | √ | P2 | 4000 | 25 | B | – | √1/11 | – |
| London (210) | √ | | 3000 | 20 | T | √ | 1/11 | – |
| London (16–19) | √ | – | 3500 | 20 | T | √ | 15/5* | – |
| Oxford Human | √ | – | 3000 | 40 | T | √ | – | – |
| Northern Ireland Physical | √ | – | 3000 | 30 | T | √ | – | – |
| WJEC | √ | – | 2500 | 20 | T | √ | – | – |

T   Teacher marked;    B   Board moderated              V   Voluntary
E   Endorsement on Certificate.                         Y   Unofficial approaches will be answered.
†   Both papers may be taken and best mark counted.     *   In Year 12 (Lower Sixth)

Table 14.1 Individual field and project work

# ESSENTIAL PRINCIPLES

Preparing a field study report is an invaluable skill because it enables you to achieve many objectives. You are expected to identify and define the purpose of your investigation, i.e. what you are studying. Having done that, you have to collect the relevant data either from primary, original sources (for example discharge measurements) and/or from secondary sources (for example published data). The information which you have gathered should then be processed, presented and interpreted using the appropriate geographical skills of cartography, statistical manipulation and representation. Above all, there should be a written explanation related to your initial objectives. To do this, the material must be presented in a clear and concise form within the prescribed length. Finally, you are expected to evaluate your findings in relation to your objectives and to be constructively critical of your own work. This all sounds straightforward, but so many studies flounder because the initial consideration of topic and theme is left too late. Field projects are not you, the candidate, versus the examination system, and you should consult your teacher/tutor at all times, especially in the early stages. Teachers know the regulations and the most successful types of study: preparation of a successful study is a partnership between you and your teacher.

## THE TIMETABLE

Most candidates who do well on the field project have allowed themselves plenty of time for the study. Assuming that the submission date is in April, then it is not too early to begin preliminary discussions by the end of your first year in the sixth form. Your actual title should be determined by October at the latest; then, with a few exceptions (such as studies of heat islands!), the fieldwork should be completed by Christmas. In this way, you give yourself time to process the data, draw those time-consuming maps and diagrams, and draft and redraft the study after consultation with your teacher. Obviously word processors (if typescript is permitted) are a considerable help in the drafting process. This whole process is tabulated in Table 14.2. If you are a one year candidate, this process must be completed between November and April.

| | | |
|---|---|---|
| June | Discuss your area of interest | |
| July | Discuss your potential topic: | Physical – Geomorphology<br>– Atmosphere<br>– Biogeography<br><br>Human – Settlement<br>– Agriculture<br>– Industry<br><br>What data is needed? |
| | Set probable hypothesis or issue or problem. | |
| Summer vacation | Examine whether it is possible.<br>(Pilot study)<br>Some boards expect you to send draft titles for approval. | – Where shall I do it?<br>Does the secondary data exist?<br>What primary data must I collect?<br>How long will it take?<br>Are there special syllabus demands? –<br>People and Environment. |
| September Year 13<br>(Upper Sixth) | Background reading: | – textbooks<br>– more advanced works<br>– journal articles<br><br>*NB* Do not copy out material. |
| October | Formulate hypothesis.<br>Decide upon issue.<br>Decide upon problem. | Get questionnaire approved. |
| October–December | Collect material, data. | Check the date to hand in. |
| January (or earlier) | Process data and draw maps, diagrams.<br>Do laboratory work.<br>Produce plan with section headings with basic ideas of contents. | |
| February | Produce final version of maps, diagrams.<br>Write first draft. | |
| March | Final production. Is the title page correct for board, with Centre Number, Candidate Number, address for return?<br>Is there a board verification of work sheet to be completed? | |
| April–June | Prepare for oral – one evening beforehand. | |

Table 14.2 A recommended project timetable

## TOPIC SELECTION

Most successful field projects take one of two forms. Either they are based on a clear, legitimate and testable hypothesis (Fig. 14.1), or they focus on a problem or issue which is readily investigated.

In both cases, they need to be precise, and this is often achieved more easily within the broad area of physical geography where rigorous testing procedures are more developed. Therefore, titles which merely state 'The development of Croydon', 'A study of the River Dane' or 'Industry in Bradford' invariably lead to broad, unstructured reports with no aim. However, titles such as 'An investigation of shop cluster in the CBD of Portsmouth', 'Factory size and distance from the CBD of Leeds' or 'Human interference with the streamflow characteristics of 5 km of the river Rother' lend themselves to more rigorous data collection and processing which can lead to a better mark.

Far too many students try to achieve too much. The study of all aspects of a CBD tends to lend itself to superficiality, whereas an attempt to define its edge could be more meaningful and manageable. Similarly, there are exercises which are not appropriate at A-level. Solving transport problems is one such topic which both governments and local authorities find difficult even with the aid of massive computer capacity. So a few imperfectly gathered counts will not fulfil your aims!

Selection of the topic will inevitably depend on your interests and the inspiration and guidance of your teacher. It might also depend on the order of study in your course. You must then be prepared to develop your ideas by reading about your chosen field of interest, first in textbooks and, perhaps in a journal such as *Geography Review, Geography* or *The Geographical Magazine*. Many of the best studies do say where they obtained their original idea and then relate it to known theories in textbooks. Thus to acknowledge studies of soil catenas from a textbook tells the examiner, for example, that you are well prepared for a study of soils, vegetation and land use on slopes of the carboniferous limestone and millstone grit of the Pennines.

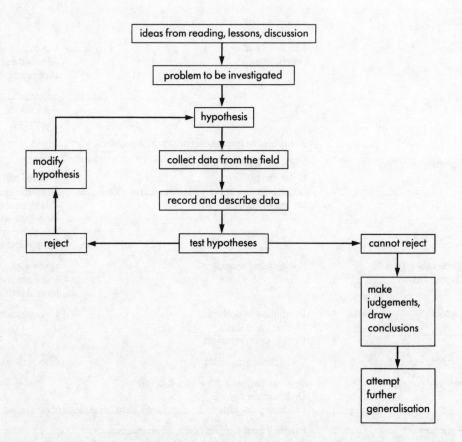

Fig. 14.1 The hypothesis testing route

Nevertheless, be wary of slavishly adhering to the recipes which you can find in many guides to field studies. While these books can provide good ideas and methods for a study, they often give you an all-too-rigid straightjacket that the examiners have seen far too often. Try to develop out from the routine discharge study, so that your study contains a new avenue of investigation.

You need to select the location(s) for your study. Once again do not be too ambitious; it is better to delimit part of a sphere of influence between two cities than to try to look at the whole sphere for one large city. Similarly, make sure that the area is suitable; it is no use testing a model of city structure against a town engulfed by suburban sprawl of a large city such as London. Is the river or slope accessible? Can you gain access to the dunes in winter? Whatever your topic, bear in mind the various stages of the 'Route for geographical Enquiry', outlined in Fig. 14.2.

## DATA COLLECTION

Before you venture into the field you need to know what data to collect. Far too often students set off to collect data and then they look for a title to fit the data. Even worse are the studies that are statistical tests in search of data. However, if you have a precise aim or hypothesis, then the type of data to be collected is obvious.

If equipment, such as a soil auger or whirling hygrometer, is to be used, then make sure that it is available and that you know how to use it. Even if the equipment is not available, then you might be able to construct your own, such as home-made rain gauges, water sampling bottles, or ready marked tape measures.

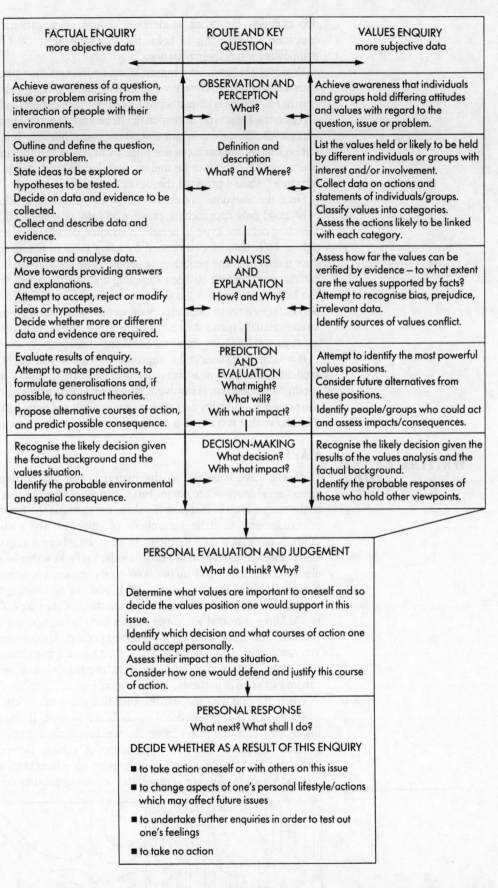

| FACTUAL ENQUIRY more objective data | ROUTE AND KEY QUESTION | VALUES ENQUIRY more subjective data |
| --- | --- | --- |
| Achieve awareness of a question, issue or problem arising from the interaction of people with their environments. | OBSERVATION AND PERCEPTION What? | Achieve awareness that individuals and groups hold differing attitudes and values with regard to the question, issue or problem. |
| Outline and define the question, issue or problem. State ideas to be explored or hypotheses to be tested. Decide on data and evidence to be collected. Collect and describe data and evidence. | Definition and description What? and Where? | List the values held or likely to be held by different individuals or groups with interest and/or involvement. Collect data on actions and statements of individuals/groups. Classify values into categories. Assess the actions likely to be linked with each category. |
| Organise and analyse data. Move towards providing answers and explanations. Attempt to accept, reject or modify ideas or hypotheses. Decide whether more or different data and evidence are required. | ANALYSIS AND EXPLANATION How? and Why? | Assess how far the values can be verified by evidence – to what extent are the values supported by facts? Attempt to recognise bias, prejudice, irrelevant data. Identify sources of values conflict. |
| Evaluate results of enquiry. Attempt to make predictions, to formulate generalisations and, if possible, to construct theories. Propose alternative courses of action, and predict possible consequence. | PREDICTION AND EVALUATION What might? What will? With what impact? | Attempt to identify the most powerful values positions. Consider future alternatives from these positions. Identify people/groups who could act and assess impacts/consequences. |
| Recognise the likely decision given the factual background and the values situation. Identify the probable environmental and spatial consequence. | DECISION-MAKING What decision? With what impact? | Recognise the likely decision given the results of the values analysis and the factual background. Identify the probable responses of those who hold other viewpoints. |

PERSONAL EVALUATION AND JUDGEMENT

What do I think? Why?

Determine what values are important to oneself and so decide the values position one would support in this issue.
Identify which decision and what courses of action one could accept personally.
Assess their impact on the situation.
Consider how one would defend and justify this course of action.

PERSONAL RESPONSE

What next? What shall I do?

DECIDE WHETHER AS A RESULT OF THIS ENQUIRY

- to take action oneself or with others on this issue
- to change aspects of one's personal lifestyle/actions which may affect future issues
- to undertake further enquiries in order to test out one's feelings
- to take no action

Fig. 14.2 The route for geographical enquiry

If you have to gather data, for example on retailing, then make sure that the classification is sufficiently detailed for your purpose. Do you need to record first-floor uses in a CBD? Try to discover the sources of data, such as timetables, delivery rounds and rate books, and make appointments if obtaining the data involves other people giving up their time for you. Once you meet these people do make sure that you know precisely what you are doing and what you need. There is nothing worse than an uncertain student with unstructured requests, and it is precisely such people who can, and do, ruin the opportunities for other students. Several organisations have already withdrawn co-operation after having to help the disorganised student who became a drain on their manpower, if not their patience.

Probably the greatest dangers in data collection face those who embark on any form of questionnaire work. Before you approach the subjects, you should pilot the questionnaire. In other words, you should write a draft of the questions and actually find out whether they produce the right type of response. Here your teacher should vet your questions to make certain that they are relevant and do not cause offence. It is even worth while trying a few out on friends and family to see if they can follow your meaning. 'Where do you go for high order goods?' might not be understood, but 'What town do you normally visit to buy new furniture?' should produce the desired response.

Once the questions are approved, you should then decide how to obtain your sample. It is no good only approaching those who *look* as if they will respond. In that way you are selecting only one type of person. You must decide whether the sample will be every fifth person or house, or determined by using random numbers to give you map references, door numbers or the sequence of people to be approached in the street. If you are in doubt, consult a text which outlines sampling and questionnaire techniques. Finally, a plea from the heart: please do not collect exactly one hundred responses because that makes the percentages easy to calculate. Such an approach is contrived and does you no credit. Good studies actually quote the response rate, i.e. the percentage who agreed to respond. Most surveys will finish up with odd numbers of responses, which is normal.

If you have to undertake any type of survey you might need the help of your fellow students. Channel measurements and pedestrian counts at a specific time work best if group data-collection is involved. However, do acknowledge all help because this increases the project's credibility and shows an awareness of data-collection problems. If well done it also shows that you are able to organise others – which obviously will appeal to employers.

## MAPS

**WRITING UP THE STUDY**

Once you have collected all the material, you will no doubt be pleased, but the most important stage is yet to come. First, you must begin to process the material. This may be in the simple form of mapping the data, for example the functions in a CBD, the newspaper advertisements to delimit a sphere of influence, the pattern of vegetation on a dune, degrees of erosion on a footpath. In all cases where a map is drawn, do make sure that it has a scale, an indication of north, a title, and a key that is suitable. Well-drawn maps are always preferable to felt-tip pen, coloured photocopies, although the skilful enhancement of a published base map can be useful. If you use an existing map then do acknowledge the source (for example, 'From D. Burtenshaw, *Cities and Towns* [Unwin Hyman, 1986], p...'). Make sure that all proper names used in the text appear on a map: this will give the examiner the impression of an organised person. Also, do ensure that the map shows what you say it does; a map of land values is not that if it uses house prices or rateable values. At this stage, write some notes on what the map shows and how you can interpret it in relation to your hypothesis, issue or problem.

When drawing your maps make sure that the symbols are appropriate and at the correct scale. Ensure that proportional symbols are located. If you are drawing choropleth maps, make sure that the class intervals are appropriate and correctly chosen and that the symbols adequately portray a gradation of values. Colour in choropleth maps can be misleading and should be used with greater care than black and white. Population gains in red and losses in blue are appropriate, but using another colour might cause confusion for

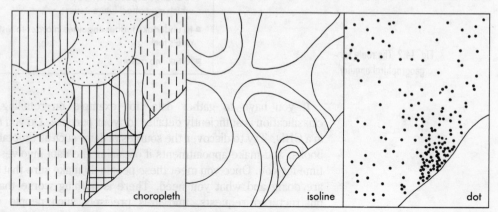

Fig. 14.3

the reader. If isoline maps are used (Fig. 14.3) then be sure that there are enough values to permit the interpolation of the line.

## GRAPHS

If data is graphable, then you should also draw as many graphs as possible, for example, stream flow and distance, percentage of heather in quadrats with attitude, temperatures along an urban transect, and then only use those which help to make your points. The scales on the graph should be chosen with care and labelled correctly. If one axis is too great, will the use of logarithmic scales help? Are the dependent and independent variables on the correct axes (Fig. 14.4 a))? If the percentage of heather cover is being related to the acidity of the soil and a variety of sites, the heather figures are the dependent variable and the acidity figures the independent variable. In constructing your graph, the data would be shown as suggested below (Fig. 14.4b)). If the graphs relating variables show a distinct pattern – linear, clustered – then it is time to ask whether this relation can be expressed statistically. Among the graphical techniques that you may wish to consider are:

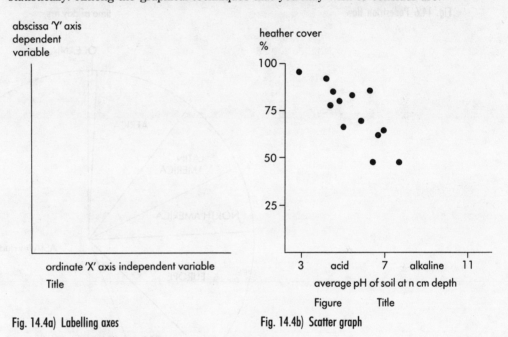

Fig. 14.4a) Labelling axes          Fig. 14.4b) Scatter graph

- **The scatter graph** (see (Fig. 14.4 b) above).
- **The line graph**, for example Fig. 14.5:

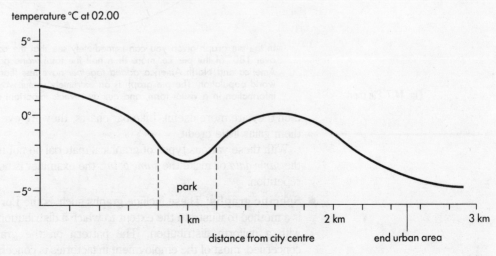

Fig. 14.5 A line graph

- **Bar charts** (Fig. 14.6): these are much overused but can make a simple point, as here, for pedestrian flow at one point through six time periods.
- **Pie charts** (Fig. 14.7): These are ideal for showing proportions, for example of different shop frontages in a shopping centre. If they are located so that, for instance, the total frontage of several shopping centres and the mix of objects can be shown,

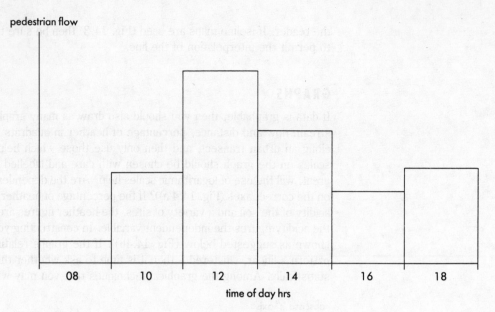

Fig. 14.6 Pedestrian flow

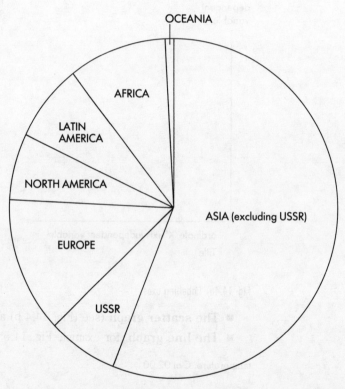

Fig. 14.7 Pie chart

In the pie graph given you can immediately see that the continent of Asia has over 180° of the pie, i.e. more than half the total world population. Africa, Latin America and North America *added together* have less than one-quarter of the world population. The pie graph is an extremely useful way of presenting information in a visual form, and quickly reveals important *relative* sizes of items.

then they are more useful. Like bar charts, they are overused and page after page of them gains little credit.

With these various types of graphical material do not use several methods to show the *same data* or make the *same point*; the examiner is not impressed by unnecessary repetition.

■ **Specific graphs**: These include graphs such as the Lorenz Curve (Fig. 14.8) which is a method to illustrate the extent to which a distribution of data is uneven compared with a uniform distribution. The pattern on the graph shows that in the city concerned, most of the employment in factories is concentrated in a few large plants.

## TABLES

These are another useful way of assembling the data which you use. However, pages of tables which are hardly used do nothing to impress the examiners. If there is a lot of basic data in tabular form, then it might be best to include a sample in the text and relegate the rest either to an appendix or to your files. As with maps and diagrams, all tables *must be used* and their salient points referred to in the text, for example: 'Figure 2.4 plots the data on floor area contained in Table 2.1'.

## PHOTOGRAPHS/SKETCHES

It might be appropriate to use photographs to aid your case. Photographs are of no value if they are not inserted in the text and used. Wherever possible they should be noted in the text, for example: e.g. 'Plates 1–10 show the meanders used in the study', and they should be titled 'Plate 1 The Field Farm Meander'. The best studies add to photographs with labels and even tracing paper overlays. Some of the more artistic among you might even care to convert the overlay into a field sketch which emphasises, for example, meander features which will be outlined in the text. Do make sure that the photographs aid your case: if you say that the study was undertaken over the summer, photographs of people doing the stream measurements on a frosty midwinter morning tend to call your statement into question. Photographs can be expensive and occasionally fail to come out, so there is no harm resorting to field sketches. No one expects you to be a landscape painter, and even the most unartistic attempts will gain credit if they are relevant to the study.

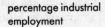

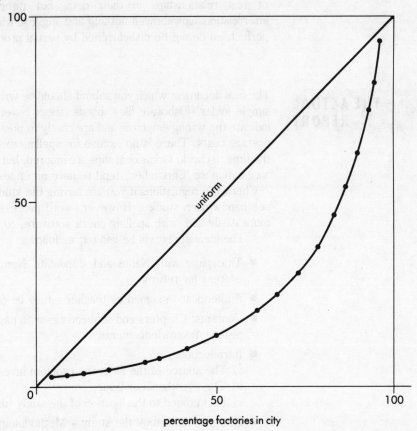

Fig. 14.8 Lorenz curve

## STATISTICAL METHODS

These are a necessary part of most geographical data processing today but they are not the be-all and end-all. Too many projects do tend to be a series of statistical tests in search of a topic. Therefore, any statistical method of description must improve the communication of your ideas to the examiner, otherwise it is not relevant. At the simplest level, percentages are used often to produce maps and graphs. More important measures of *central tendency* or the average, mode, median or mean can be used. If these are used, then they should be accompanied by measures of spread or variation, the standard deviation and the range. Means and standard deviations are most useful for deciding class boundaries on choropleth maps.

The methods which you might use in your study can obviously be chosen from the list which your board includes with its syllabus. Many boards even give the versions of the formulae that they use and so you could use that. For example, WJEC A/S lists mean, standard deviation, standard error of the mean, nearest neighbour index, chi square, Spearman rank correlation and location quotient. Other boards will note more: comparative tests such as Mann-Whitney U test and Student's t test or relationship tests such as Pearson's product–moment correlation and even the testing of trends through linear regression. If you use any test, make sure that the examiner is given a reference to the source text for your chosen method (for example: D. Ebdon, *Statistics in Geography* (Blackwell, 1977), and that you are aware of what it shows, especially if you have an oral examination (see Table 14.1). If you do use correlations and comparisons, then there are tests for significance which should also be carried out to give completeness.

There is only a need to show the working of a test once in your report. After that you can just state, for example, that 'further Spearman rank tests were carried out on distance from supermarket and frequency of visit and distance, and age of respondents, and that these gave results of $R_S$ 0.81 and 0.90 which were significant at the 5 per cent level'. Here, as always, the data has been processed; it is now up to you to interpret those results in relation to your hypothesis or problem. If they show what you hoped, then you reaffirm your hypothesis or solve the problem. Do not be discouraged if the results are not what you expected. First, test the data again and, second, ask or check to see if you are using the correct procedure. If the result stands, then ask yourself why it is so and put this in your analysis. Similarly, if several draft graphs show little correlation, do not redraw them or test relationships in their data, but rather state that an examination of the interrelationships showed nothing and suggest why that was so. Most findings are never perfect, so do not be disheartened by partial proof or rejection of an hypothesis.

## THE ACTUAL REPORT

The final document which you submit should be written in a clear fashion and presented in a simple folder. Elaborate files, plastic sheet covers and sticking pages on card normally indicate the wrong emphasis and are costly to post. Some boards will remove these to cut postage costs. There is no excuse for spelling mistakes in a report because you do have the time to check. Occasional slips are ignored, but persistent misspellings of, for example, 'vegetation' or 'Christaller', tend to have an undesirable impact.

Check the regulations if you are having the study typed, because some boards wish to see handwritten studies. However, word processing does enable you to revise a study more easily and, with spelling check software, to eliminate most errors.

The ideal study will be laid out as follows:

- Title page with Name and Candidate Number, Centre Name and Number, and address for return.
- Authentication signed by teacher – may be on title page.
- Contents: Chapters and subheadings with page numbers. Lists of figures, tables and plates. Acknowledgements.
- Introduction:
  a) The source of the idea, the problem investigated, leading to
  b) The hypothesis or issue.
  c) Background to the location of the study (the examiner might not know the area).
- How you undertook the study – Methodology.
  Data sources.
  When collected, how collected. Samples. How recorded.
  Questionnaire should be included.
- Analysis and interpretation of data in relation to hypothesis or issue.
- Conclusion and critical evaluation of your study.
- Bibliography: books, articles, maps and other sources, for example, newspapers.
- Appendixes – if needed, but try not to use.

On the other hand, it is possible to use the Route for Geographical Enquiry laid out in Fig. 14.2, as the structure of a project. This is recommended for those taking either London 16–19 A-level or its A/S equivalent.

Do make a copy of your work – the Post Office have been known to lose projects!

**ORAL EXAMINATIONS**

Both Cambridge, and Oxford and Cambridge project papers have an oral examination on the basis of your submitted study. This examination lasts for approximately 15 minutes and its purpose is to give examiners the opportunity to be convinced by the mark which has been awarded. The examiner is only entitled to move the mark slightly in either direction and downward movement of marks has to be really justified.

The examiner is looking for proof that you understood what you did and can explain your results. Some examiners will ask to see your data records and notes, so take them along. The format will depend on your topic but you should be able to answer questions on the following:

1 Why you chose the topic;
2 The techniques which you used, their advantages and drawbacks;
3 The reasons for your conclusions;
4 Any improvements that you would make with the benefit of hindsight.

Therefore you should re-read your copy of the field report before the oral examination so that you are familiar with work which you completed maybe two or three months beforehand.

**POSSIBLE PROJECT TITLES**

### Some boards set field titles for AS-level

These titles are from the AEB 1988 syllabus:

■ Test the relationship between soils, geology and human activity in a selected area.

■ Examine the role of people in altering the hydrological characteristics of a 5 km stretch of a stream.

■ Test the hypothesis that an urban area modifies the pattern of weather experienced by a town or city.

■ Test the hypothesis that the limits to the sphere of influence of two towns form a zone rather than a line.

■ Examine the changing patterns of industrial location in a selected area or city.

### Issues suitable for A-level and AS-level reports

■ The deregulation of buses and its effects on public transport in North Wales.

■ The geographical implications of by-passing a town or village.

■ The effects of out-of-town retailing on Southampton.

■ The problems of waste disposal in Tyne and Wear.

■ The policy for river valley recreational areas in Greater Manchester.

■ The effects of extensive public use in parts of the Pennine way.

■ The causes, effects and possible solutions to the problem of coastal erosion along a (named) stretch of coastline.

■ The proposals to build on an area of Birmingham's Green Belt. Who gains, loses and why.

### A-level project titles from the past – with a brief comment (topics in the two previous sections would also be acceptable.)

■ The sphere of influence of Clacton-on-Sea.

■ The sphere of influence of Limavady.

*Routine topics which need good survey data.*

■ The Central Business District of Reading.

– *Vague and probably too big a topic.*

■ Retail affinities in Central Portsmouth.

■ Is there a core–frame in Southampton?

*Good topics amenable to testing.*

■ Does Kingston-on-Thames conform to the Burgess model?

– *Poor, vague topic, too big.*

- Ethnic areas in Bedford – the use of electoral rolls.   – *Original topic.*

- Testing Christaller in rural Essex.   – *Potentially good; much travel needed.*

- Industry in Bracknell.   – *Too vague and generalised.*

- Industrial type and distance from the centre of Newport.   – *Good topic.*

- Can we solve Birmingham's traffic problem?   – *Far too big and unwieldy.*

- Soils, vegetation, agriculture and geology on the North Down escarpment.   – *Precise topic amenable to testing.*

- The effects of settlements on water quality on the Dee.   – *Good overarching topic – needs good laboratory skills.*

- The changing population structure of villages with distance from York.   – *Good topic needing secondary and primary sources; perhaps big.*

- The catchment of a new superstore.   – *An overworked topic.*

- The catchment of Coleraine Sports Centre.   – *The topic of the sports enthusiast; needs good data, otherwise thin.*

- The social geography of Ocean Village.   – *Depends on good sample data; difficult but original.*

- Land use and distance from the farm in East Anglia and Central Wales.   – *Good idea but difficult to ensure that both parts of equal standing.*

- The downslope movement of fertilisers.   – *Original but needing good laboratory facilities and field skills.*

# INDEX